Coral Empire

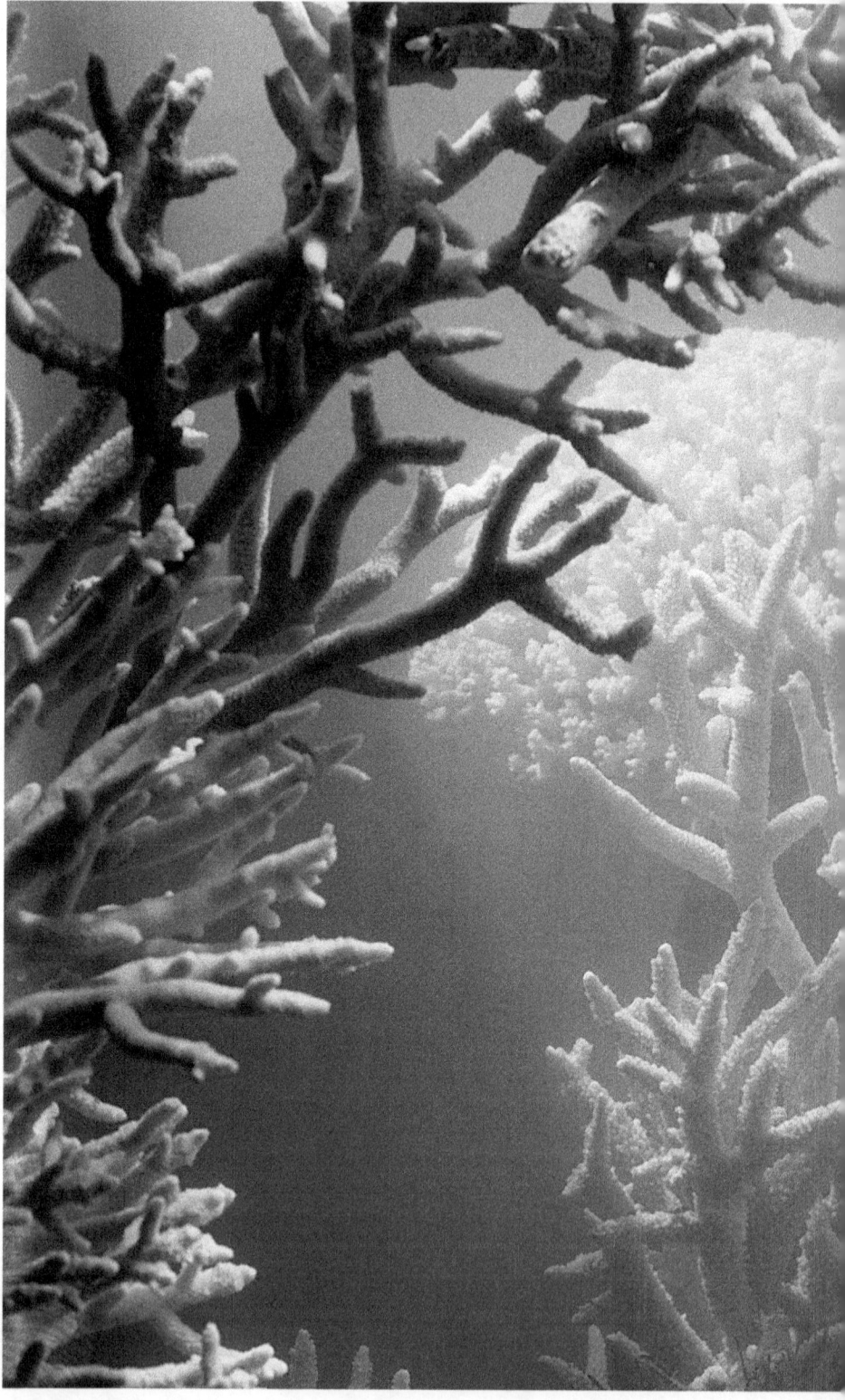

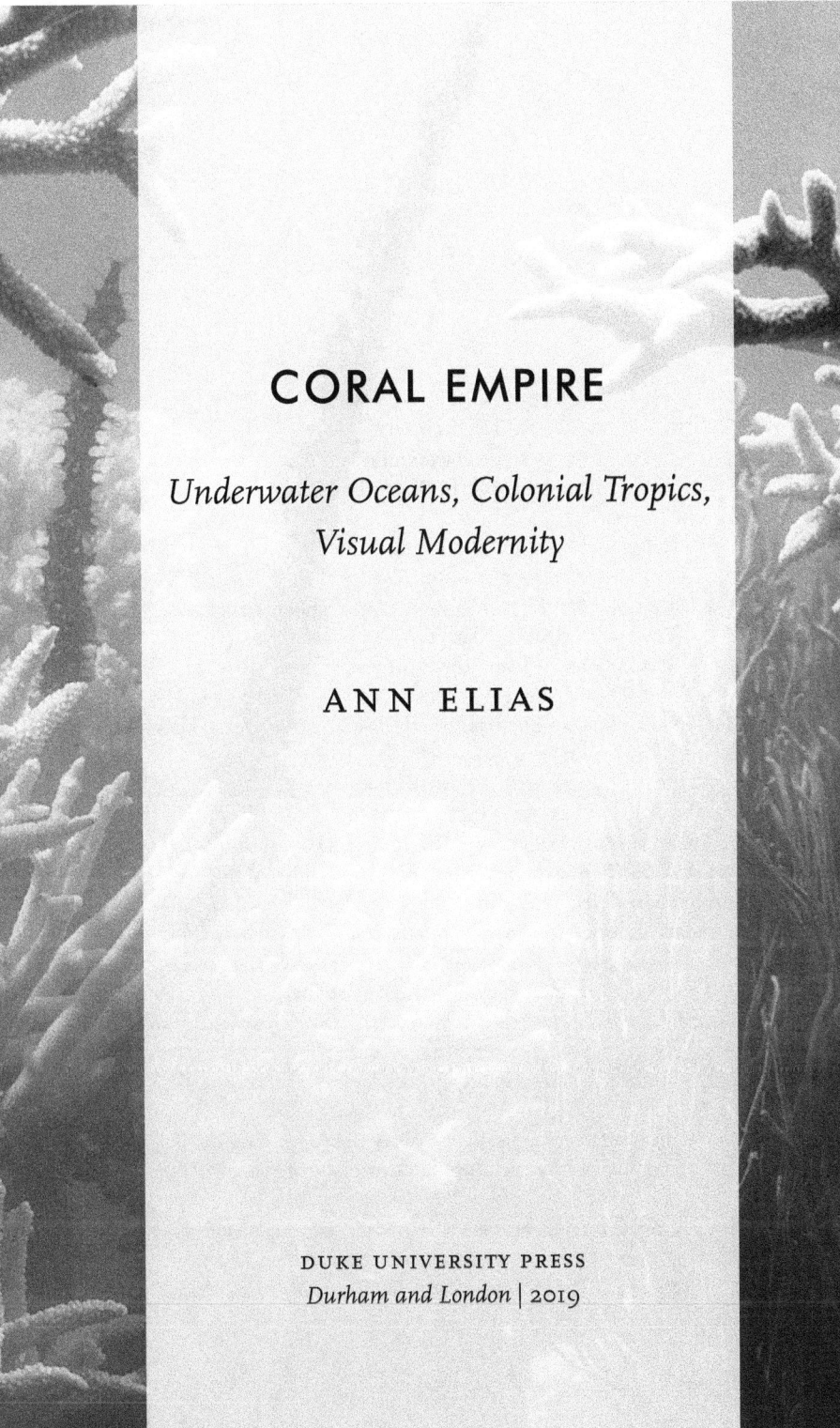

CORAL EMPIRE

Underwater Oceans, Colonial Tropics,
Visual Modernity

ANN ELIAS

DUKE UNIVERSITY PRESS
Durham and London | 2019

Designed by Jennifer Hill
Typeset in Scala by Copperline Books

Library of Congress Cataloging-in-Publication Data
Names: Elias, Ann, [date] author.
Title: Coral empire : underwater oceans, colonial tropics,
visual modernity / Ann Elias.
Description: Durham : Duke University Press, 2019. |
Includes bibliographical references and index.
Identifiers: LCCN 2018037352 (print) | LCCN 2018057700 (ebook)
ISBN 9781478004462 (ebook)
ISBN 9781478003182 (hardcover)
ISBN 9781478003823 (pbk.)
Subjects: LCSH: Williamson, J. E. (John Ernest), 1881-1966. |
Hurley, Frank, 1885–1962. | Coral reefs and islands—Research. | Underwater
exploration—History—20th century. | Underwater exploration—
Environmental aspects. | Underwater photography—History—20th century. |
Underwater photography—Social aspects. | Ethnology—Social aspects. | Visual
anthropology. | Other (Philosophy)
Classification: LCC QE565 (ebook) | LCC QE565 .E45 2019 (print) |
DDC 770.092/2—dc23
LC record available at https://lccn.loc.gov/2018037352

This project was assisted by Sydney College of the Arts,
the Visual Arts Faculty of the University of Sydney.

Cover art: Frank Hurley, "A harmonious group of soft
and stony corals seen through the transparent water," ca. 1950.
From Frank Hurley, *Australia: A Camera Study* (1955), 94.

To Greg, with love

CONTENTS

ACKNOWLEDGMENTS

This project was enabled by a Harold White Fellowship at the National Library of Australia in 2014. Thank you to the library staff at the NLA for access to collections and images.

Many libraries and museums were generous with collections and images, but without the cooperation and generosity of Armand Esai and the Field Museum Archives, Patricia Egan and the Australian Museum Archives, and Constance Krebs and the Association Atelier André Breton, this book would never have taken shape. And this applies also to the generosity of artists Peter Peryer and Frederico Câmara.

I greatly appreciate the expert, critical feedback and guidance from the book's two readers, and the interest and care taken by editor Courtney Berger, editorial assistant Sandra Korn, and project editor Sara Leone at Duke University Press.

Colleagues and students at the University of Sydney make up the creative community that nurtured this research: Iain McCalman, director of the Sydney Environment Institute, whose extensive work on the Great Barrier Reef has been an inspiration; Robert Dixon, in the English Department, who shared his deep knowledge of Frank Hurley's life, work, and myth; artist and writer Adam Geczy who helped me think more creatively about images in this book and broadened my perspectives on art and visual culture. Thank you to artist Helen Hyatt-Johnson, an alumna of the University of Sydney, for reading the third draft, for conveying her perspectives on the book, and for her support and enthusiasm. Dominica Lowe, librarian at Sydney College of the Arts library, helped me by tracking down sometimes-obscure films, videos, books, and articles. And Michelle St Anne, deputy director of the Sydney Environment Institute enabled me to become part of an exciting group of humanities scholars and scientists at the University of Sydney.

Thank you to Martyn Jolly, my colleague at the Australian National University, Canberra, for explaining how magic lanterns work and for sharing information and ideas about photography and Frank Hurley. And to Randy Olson, who formatted the book, organized references, and completed the bibliography and index.

More than anything I want to acknowledge my families in New Zealand and Australia. A special thank you for a lifetime's support to my mother, Mair, my late father, Minas, my sisters, Sian and Victoria, and my brother, John, and for the books, articles, and ideas they put my way, many of which are contained in this study. Greg, my partner, and my daughter, Rose, were there to experience the surreal and beautiful corals, rays, epaulette sharks, reef sharks, sea cucumbers and turtles at Heron Island at the Great Barrier Reef in 2014; their conversations and our experiences have contributed greatly to this project.

Parts of chapter 2 appeared in Ann Elias. "Sea of Dreams: André Breton and the Great Barrier Reef." *Papers of Surrealism*, no. 10 (2013): 1–15. Parts of chapters 3 and 4 appeared in Ann Elias. "Second Life: Chicago's Bahama Islands diorama." *Antennae: the journal of nature in culture*, no. 29 (2014): 34–49, and Ann Elias. "Ocean Monster Screen Stars: The impact of J.E. Williamson in Australia." *Australasian Journal of Popular Culture*, no. 4 (2015): 85–99.

INTRODUCTION

Over centuries, the coral reef has figured as a mariner's nightmare, a scientific problem, the source of myth, a visual object, a touristic landmark, an Indigenous heritage, and, for explorers, an underwater frontier. This investigation focuses on the emerging compulsion in the early twentieth century to photograph and film coral reefs underwater, to capture still and moving imagery of tropical marine life in the wild, and to present the results to the public as a brand new photographic and cinematic experience. The overarching argument of this study is that by the 1920s, mass media culture, together with new technologies related to cinema, photography, and museum exhibitions, produced coral reefs and the underwater as a modern spectacle. But while the book's argument as a whole is organized around the spectacle of the reef and the underwater, it also foregrounds three themes that recur as points of discussion throughout chapters: the mediation of vision of the underwater and manipulation of truth by modern technologies; the racialization of coral reef environments and subordination of coastal peoples in colonial modernity; and the objectification of marine animals as source of knowledge and entertainment.

Of all the oceanic zones, the underwater is the least discussed region, particularly for the period of most concern to this study, the 1920s, and that decade's wider cultural context, the period from 1890 to 1940. Across that time span, the aura that had long surrounded the underwater—the earth's most mysterious and invisible domain—was

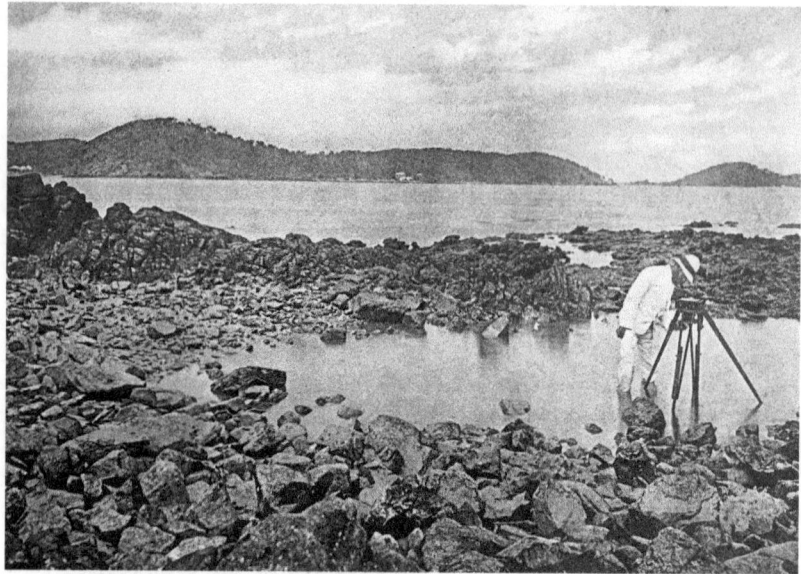

Fig. I.1. William Saville-Kent photographing the Great Barrier Reef.
From W. S. Kent, *Great Barrier Reef of Australia*, 1893, 183.

reignited in three realms of visual culture. In science, there was a push to study fish and undersea plants in their native states with cameras submerged.[1] In avant-garde circles, artists such as Paul Klee, André Kertész, and André Breton found new aesthetic and symbolic meanings in the concept of realities that lie submerged beneath the surfaces of things, especially water. And in the realm of popular culture, where general audiences sought escape from everyday life, photographers and cinematographers sensed that in the underwater lay a profitable new beginning for images and imaginings. In all realms of science, art, and popular culture, the underwater promised an expansion of vision through representations of sights normally unseen.

The underwater, a wild zone at the margins of the modernizing world, was the newest possible viewpoint on the planet and promised filmmakers and photographers of the early twentieth century a rebirth of visual culture that would put their careers at the center of a modern triumph. There was a perception around 1910 that planetary space was shrinking, wilderness diminishing, and adventures for explorers dwindling. But ... there still remained the ocean, and the ocean's underwater. Westerners, who were outsiders to the undersea, perceived it as a blank space, an uninhabited space, an unlived space belonging to no one and nothing. They saw it as something waiting to be filled. The idea of penetrating that tabula rasa with cameras was irresistible.

Coral Empire explores Western visual culture in the early twentieth century in relation to the urge to make history by photographing and filming the underwater, and it does this by investigating two very specific geographical locations: the Bahamas in the Atlantic, and the Australian Great Barrier Reef in the Pacific. The narrative structure of the book is based on the stories of two men who were filmmakers, photographers, and explorers of renown. They almost certainly never met, but in the 1920s, in the name of science, popular culture, and exploration, their aim was to leave their mark on the world by the visual conquest of the tropical underwater. *Coral Empire* is about the striving, the longing, the solutions, and the failures of these men in meeting their objectives.

At the Bahamas and the Tropic of Cancer was John Ernest Williamson (1881–1966), known variously as "Jack Williamson," "Ernie," "J. E.," and "J. E. Williamson," who photographed in a marine paradise of transparent water and white sand in a country of scattered islands and cays where the undersea, at its best, was, as he put it, a realm "of marvelous things, all set in the palest of sapphire mists."[2] At the Tropic of Capricorn was James Francis Hurley (1885–1962), known as "Frank Hurley" and "Captain Frank Hurley," working in the region of the Great Barrier Reef, a place known as a miracle of nature and the world's largest and most beautiful underwater coral realm, with a reach extending from the Coral Sea off the coast of Queensland through the island chains of the Torres Strait into the Gulf of Papua.[3] The tropical water surrounding the coral reefs of the Australian region was sometimes so clear and so like crystal that Hurley described it as "liquid glass."[4]

By the 1920s, the coral reef had become a space of imagination in a different way. Stefan Helmreich explains the difference as a shift from perceiving reefs as structures and architecture to appreciating them as voluptuous environments. In the 1920s, it was the living environment of reefs that people sought; Helmreich describes it as a desire "to submerge their bodies and eyes in the midst of" living corals.[5] This is also what Hurley and Williamson sought. The prized object for their photography was the sea floor. The prospect of photographing and filming coral reefs and submarine life eye to eye symbolized their modernity and brought their project close to their hero, Jules Verne (1828–1905). It was Verne who gave Hurley and Williamson a consciousness of the sea as an alien outpost of the world, and one awaiting discovery, and Verne who attracted so many filmmakers to the underwater.[6] It was also the impact of Jules Verne on Western perceptions of the undersea that explains why both Williamson and Hurley were referred to in their lifetimes as "Captain Nemo."

But Hurley and Williamson were not the first to want to explore the seafloor or the underwater at coral reefs, nor the first to try to capture those sights photographically. Throughout the nineteenth century, photographers had longed to do this, including the British scientist William Saville-Kent (1845–1908). It was relatively straightforward to take a viewpoint of a coral reef by looking down at it, vertically, as seen in figure I.1, a photograph of Saville-Kent at work in the 1890s, recording marine life at the Great Barrier Reef of Australia. But the limitations for visualizing the underwater are obvious. The viewpoint is distancing and abstracting. Imagine a fish photographed from above emerging in the image as a vague shape without face or eyes or sense of presence to the viewer. Only from a viewpoint underwater could people also understand the sea from within it rather than from outside it. That is why the shift from land and air camera viewpoints to perspectives beneath the surface of the water changed the course of history. It altered science, influenced tourism, broadened popular knowledge, and eventually, in the late twentieth century, helped bring about greater ecological understanding of the lives of marine animals and plants and established traditional peoples' custodianship of seabeds in public consciousness.

By 1910, only a handful of people, mostly scientists, had managed to photograph the underwater and obtain images, among them the French zoologist Louis Boutan (1859–1934).[7] The attempts these men made at underwater photography were groundbreaking. But Hurley and Williamson wanted something more. Their ambition was to supply an international public with startling visions of the tropics that the public had previously sought through literature and stories. The medium of film, moving and still, suited them perfectly. It promised literal realism but was easily fictionalized and manipulated.

As the following chapters reveal, Hurley and Williamson benefited from the belief system that "seeing is believing." In their ambitions to turn the seafloor and the underwater into entertaining spectacles, they bent the truth. Consequently, *Coral Empire* argues that in relation to the developing culture of spectacle created by mass media, the virtual world of cinema, and the global circulation of photographic reproductions, Hurley and Williamson were producers of a culture of the copy in which it was increasingly difficult to distinguish reality from truth. Yet they marketed their work on claims of authenticity and realism. In the hyperreal world of visual media in which they operated, photographic images were frequently detached from original contexts, geographies were often misrepresented with inaccurate captions, and deceptions were commonly practiced but concealed in the interests of entertaining documentaries about the real, but mysterious, natural world. The problem of "reality"; the "authenticity" of places, people, and events; and the collapse of borders between reality and fiction, emerges, then, as a key consideration in the lives and works of Frank Hurley and J. E. Williamson.

In fact, this book's very conception was a response to the curious circumstances surrounding the authenticity of an underwater photograph of a coral reef. The image in question is reproduced in *Mad Love* (*L'Amour fou*), a book published in 1937 by André Breton (1896–1966), the leader of the surrealists. Below the image a caption reads, "The Treasure Bridge of the Australian Great Barrier," and attributes the image to the *New York Times*. The photograph is conspicuous because the number of images of the Great Barrier Reef taken underwater and published in the 1930s was drastically few.[8] In the process of researching the image, I discovered that the photographer was J. E. Williamson, and the photograph was a picture of the Bahamas underwater, not the Australian Great Barrier. From there, *Coral Empire* became an inquiry into commonalities and overlaps in the histories of how photography and film made the coral reefs and the underwater of the Bahamas and the Great Barrier Reef modern marvels. The book also became a study of what happens to the meaning of an image when a photograph circulating in mass media is removed from its original context and placed in a new one.

In the context of the history of early cinema, Hurley and Williamson worked between two paradigms: the era when performers or showmen used short films as props for theatrical acts in which lectures and lantern slides were also integrated, and the era of feature films.[9] Their audiences were general, but also from the fields of science and exploration. Both men toured the United States, Britain, Australia, and Canada speaking and showing films and lantern slides of coral reefs, native peoples, and the tropics in public venues such as the Smithsonian Institute, Carnegie

Hall, the Field Museum, and the American Natural History Museum. They employed agents to market their films, organize lecture circuits, and promote their work. They were image hunters. The way they approached the sea and the underwater was perfectly characterized by Susan Sontag, who explained the psychology behind the increasing impact of photography on modern life as "that mentality which looks at the world as a set of potential photographs."[10]

The act of diving features in all the stories told here. Williamson was a practiced suit-and-helmet diver; Hurley claimed to be practiced at diving, but the proof is hard to find. Regardless of their ability to dive underwater with suits or without, there was little or no capacity in the early 1920s to work with cameras and the body immersed in the underwater, even for short periods of time. How they got around that problem is eccentric and intriguing. It makes their stories absorbing. It is also the source of a key argument threaded through this study: that the way Hurley and Williamson conceptualized the undersea and visualized marine animals was mediated by the artificial underwater environments of aquariums. The aquarium is an optical device invented for viewing the underwater and its creatures in the safe, dry space of land and air. It emerged as a popular form of public and private amusement in the nineteenth century. Aquariums magnify the view beyond and produce the illusion of closeness to the underwater but maintain a rigid separation between human and nonhuman life. The implications of aquarium thinking in relation to J. E. Williamson have recently been elaborated by Jonathan C. Crylen. In "The Cinematic Aquarium: A History of Undersea Film" (2015) Crylen concentrates on how J. E. Williamson conceived of the sea as an aquarium, and offers the insight that cinematic vision of the underwater was indebted to discourse around this technology.[11] As I show, Hurley and Williamson both imagined the ocean as a vast aquarium, and also as a gigantic optical device for the projection of light effects. Hurley, for example, noticed how water sent beams of light through the liquid medium in a way that recalled how rays of light are projected through the thick air of a movie theater.[12]

Coral Empire reflects a shift in scholarship from land to sea. Steve Mentz and Martha Elena Rojas name this cultural turn "an ocean-inflected 'blue humanities'" in honor of "the largest and most alien spaces on our blue planet."[13] Scholars are paying more attention to the importance of ocean environments and the underwater to cultural, literary, technological, scientific, and environmental histories. Margaret Cohen, in "Underwater Optics as Symbolic Form," looks at early twentieth-century representations of the undersea and concentrates on a time in European history when the submarine realm was conceived as a new planetary space. She argues that one condition of modernity was the way "technologies enable new modes of perception, which transform the imagination and inspire the arts. A vivid example of this process is the transvaluation of the underwater environment in the Western imaginary."[14] But, as Cohen also explains, visualizations of the underwater were persistently constructed in the modern period through old conventions of linear perspective that contradicted the nature of subaquatic optics in which colors and forms behave very differently than those of land and air. What, she asks, did representations of the underwater world at that time reveal to the general public about submarine reality?

Coral Empire takes up the challenge of investigating early twentieth-century visual representations of the underwater realm and their public reception in the West. As a history of visual culture relating to coral reefs at the Bahamas and the Australian Great Barrier Reef—two outposts of the British Empire—the book is also a study of the type of imagery that Paul Gilroy terms "imperial phantasmagoria"—dazzling and symbolic artifacts that include illustrated magazines, magic lantern slides, dioramas, aquariums, travelogues, and wildlife films, all informed by and consumed amid social, cultural, and political circumstances relating to empire.[15] When the "Age of Empire" drew to a close at the time of the First World War, as E. J. Hobsbawm claims, it had created an immense visual record of its work in the form of advertising, illustrated periodicals, and still and moving images that gave those at home some insight into the remote, the unseen, and the imagined world "out there" in the colonies.[16]

To anyone who was British-descended and white, the coral islands, reefs, and waterways of the Bahamas and Australia in the 1920s were known as the empire's "possessions." The "Edenic isles set in sparkling seas"—as David Arnold described Western ideals of the tropics—generated much imperial self-satisfaction.[17] With these exotic seas and islands in its possession, the British Empire was also a "coral empire" in which the figure of the coral reef became a suggestive symbol of expansionism. It gave the empire a framework for understanding itself. Symbolically, the reef, with its busy, colonizing "workers" and expanding territory, came to define the imperial project, and in that scenario the body of the reef was imaginatively mapped onto the figurative body of Britain.

But not everyone found the acquisition of coral islands a positive direction for the expansion of empire, seeing instead the accumulation of useless lands and peoples, and dangerous environments.[18] The problem of "systematic race-thinking" is something Paul Gilroy puts down to the way "truths" about race have been constructed through Western knowledge and power.[19] Without doubt, the beauty of coral islands and their cultural desirability stand in stark contrast to the uglier realities of colonialism and racism that mark the histories of the tropics. At the center of British colonial influence in the Bahamas and the Great Barrier Reef were groups of maritime peoples—descendants of Africans, people indigenous to coral islands, and descendants of Pacific Island peoples. Today some call themselves "black," others "Indigenous," and others by the name of the islands they live on, such as Torres Strait Islander.[20] In the Bahamas, they were connected to Britain through colonization in the eighteenth century, and throughout the history of slavery. The peoples of the Pacific who are discussed here, however, were colonized by the British Empire but became subjects of Australian colonial administration. The social principles were global: mobile, white explorers and adventurers benefited from knowledge of the sea acquired from maritime peoples, especially skilled divers known ubiquitously by the colonial term "native divers," whose knowledge was passed down over generations but who were treated as a servant class employed to further European knowledge and progress, and were never properly acknowledged. Many of the images in this book show how black and Indigenous peoples were either pushed to the background or turned into spectacles by Frank Hurley and J. E. Williamson.

What kind of people, and colonials, were Hurley and Williamson? What has been said about Hurley by A. F. Pike also applies to Williamson: these men were self-styled loners "who braved danger in exotic areas to provide romance and adventure for armchair travellers" of the tropical colonies of the British Empire.[21] Typical codes of masculinity in the period meant Hurley and Williamson both wore paramilitary khaki and tropical whites, valorized "bravery, fearlessness, physical fitness and strength," and admired a man's aptitude for technology and engineering.[22] They were the type of explorer that Felix Driver called "the foot-soldier of geography's empire."[23] They were dedicated to expanding the reach of the Western world into territories they imagined it was their right to take.

Both men were born in the late nineteenth century, when oceans were portrayed in literature as mutable forces: sometimes motherly, beautiful, and gentle, and other times the embodiment of evil. Williamson, for example, was consumed by what he saw as the ugliness of sharks, and devoted years of his life to filming the horror. In fact, animals, dead and alive, in aquariums and dioramas, in natural history museums and in the field, in photographs and on film, are central to this story of underwater exploration and representation of tropical coral reefs. Consequently, the question of "the animal" emerges in *Coral Empire* as a central point of discussion. Fish in tropical waters, for example, were by turns objects of beauty, curiosities, problems, and adversaries. To further knowledge of the natural history of the sea and its animals, Hurley and Williamson worked closely with some of the Western world's most prominent museums. But their work often involved destruction of the natural environment. While the animal as camera subject was vital to underwater filming and photography, the animal was often a victim. The stories of Hurley and Williamson expose how the desire to look at animals, to hunt with cameras, and to consume the exotic world through photographic reproductions and cinematic projections embodied symbolic as well as physical violence.

In addition to archives comprising letters, papers, films, photographs, and newspaper articles, a wide range of literature has informed this study. Some authors have been cited already, but special mention should be made of others. In more than one publication, but in particular *Photography, Early Cinema, and Colonial Modernity: Frank Hurley's Synchronized Lecture Entertainments* (2013), Robert Dixon explores in detail how Frank Hurley was shaped by modern times and by colonialism, and elaborates on Hurley's central place in the spectacle of empire. Krista A. Thompson puts J. E. Williamson in social and cultural context in *An Eye for the Tropics: Tourism, Photography, and Framing the Caribbean Picturesque* (2006), in which she explains the segregated racial environment of the colonial Bahamas where Williamson's identity was shaped. In *The Reef: A Passionate History* (2013), Iain McCalman devotes an entire study to the power of corals at the Great Barrier, covering the periods from the "discovery" by captain James Cook to environmental activism to stop the destructive impact of mining and industry in the late twentieth century.[24] In the expanding literature on oceans and the underwater, Natascha Adamowsky, in *The Mysterious Science of the Sea, 1775–1943* (2015), explains how the frontier status of the underwater in the modern period was informed by aesthetic wonder as well as science. Margaret Cohen, in *The Novel and the Sea* (2010), details the significance of the sea to international literary history. And Helen M.

Rozwadowski, in *Fathoming the Ocean: The Discovery and Exploration of the Deep Sea* (2005), unlocks the history of ocean explorers and the development of technologies for acquiring knowledge about the deep.[25]

The question of cinema, photography, and the animal is explored by Jonathan Burt in *Animals in Film* (2002), in which he demonstrates the symbiotic relationship between the development of film technology and the animal as object of study.[26] The animal is also addressed by John Miller in *Empire and the Animal Body: Violence, Identity, and Ecology in Victorian Adventure Fiction* (2014), in which Miller fleshes out the animal's relations to empire and colonialism and the racialized borders between human and animal.[27] And Carrie Rohman, in *Stalking the Subject: Modernism and the Animal* (2009), offers insight into the anxieties and their consequences evoked by Charles Darwin's theory that there is no border between human and animal.[28] There are very few texts on early cinema and the underwater, but, in addition to Crylen's aforementioned thesis, Nicole Starosielski's "Beyond Fluidity: A Cultural History of Cinema under Water" (2013) identifies three phases of underwater filmmaking. Her conclusions about the first phase of filmmaking from 1914 to 1930 correspond with the findings of this investigation, namely that in this period the seafloor was an object of desire, and the underwater a place conceived by whites as the space of the racial Other who was also imagined as part of the fauna and flora.[29]

Part I of *Coral Empire*, "The Coral Uncanny," begins with a chapter that looks at the significance of coral reefs to modern visual culture and how the popular imagination of the coral reef was created. Chapter 1 is where the book's title, *Coral Empire*, is given context, and it explains how coral reefs became privileged objects of the Western imaginary. It is followed in chapter 2 by an account of how the image of the coral reef is taken up in the 1920s and 1930s by the European avant-garde, concentrating on the example of André Breton and drawing connections between surrealism, underwater space, and coral reefs. I have already mentioned the underwater photograph of a reef reproduced in 1937 by André Breton in *Mad Love* captioned "The Treasure Bridge of the Australian Great Barrier"—it serves as a point of entry to the history of underwater photography and cinema and their public reception, to the stories of Frank Hurley and J. E. Williamson, and to why *Coral Empire* looks at the Bahamas and the Great Barrier Reef as a single study.

Providing the biographical and professional background of J. E. Williamson is the purpose of part II, in which the chapters focus on work he undertook in the 1920s when hunting, capturing, and filming marine animals and corals for dioramas destined for the American Museum of Natural History in New York and the Field Museum in Chicago. Chapter 6 of this section provides a critical link to Frank Hurley, who then becomes the subject of part III. This section on Hurley concentrates on two scientific expeditions: the first to the Great Barrier Reef, the Torres Strait, and Papua in 1921; the second to the same region during a collaboration in 1922 with the Australian Museum, Sydney. From these expeditions, Hurley produced *Pearls and Savages* (1921, with a later iteration in 1923), a film for general audiences, notable for scenes of coral reefs that are almost certainly the first film footage of the Great Barrier Reef depicted for an international public.

Following individualized studies of Hurley and Williamson, parts IV and V investigate commonalities and connections between them. One chapter scrutinizes them in relation to explorers, the Explorers Club in New York, and Carl Akeley (1864–1926), celebrated curator of the American Museum. The chapter that follows then assesses Hurley's and Williamson's status and engagement with the underwater in the 1950s during a technological paradigm shift toward mobility and the immersion of photographers in the underwater after the commercialization of tropical destinations saw an explosion of tourism.[30] The chapters end with Part V, in which Hurley and Williamson are considered in the context of the Anthropocene and the acceleration of anthropogenic impact on the planet's coral reefs. This section is informed by wide recognition of the extensive and often irreparable damage from global warming to the world's reefs, particularly the Great Barrier Reef and the reefs of the Bahamas.[31]

Coral Empire is a study of a period before, but yoked to, the planetary challenges faced by the contemporary ecological crisis and by the continued impact of colonial modernity on the island peoples of the Bahamas and Australia. Indeed, issues of environmental damage and racism addressed in this investigation are intertwined in a way that relates directly to Ghassan Hage's argument that ecological struggles and racial struggles unfold together in history: they are fundamentally linked through a Western disposition to dominate and exploit the Other.[32] The ugly reality of racism, as it relates to the stories told here, is manifest in the way black and Indigenous peoples were exploited through labor and characterized as inferior and sometimes as subhuman. Related to human exploitation was environmental exploitation that resulted in unsustainable quantities of shells, pearls, sponges, corals, and fish extracted from the Bahamas and the Great Barrier Reef. However, following the period addressed in this book there occurred unprecedented development of coral reef environments, not only through extraction industries and mining but also through tourism and the profitable spectacle that coral reef environments promised. It was tourism that made coral reefs the most fetishized spaces of the oceanic environment, portraying them as outside civilization, as untouched by human culture, and as the most colorful stages imaginable for human players. Eventually, environmental impact would transform the reefs of the Bahamas and the Great Barrier from wondrous to endangered, and in many places from ecologies of vibrant color to what Jeffrey J. Cohen terms the "inhuman" color gray.[33] In an age of mass bleaching, the world reads the grayness of bleached coral as the melancholic sign of a dying planet. But in the modern period, before reefs turned gray, photographers and filmmakers, as well as museum designers, brought delight to audiences who would never experience a coral reef themselves by allowing them to escape to the tropics through the immersive effects of photography and cinema, and the cinematic effects of life-size coral reef dioramas.

PART I

The Coral Uncanny

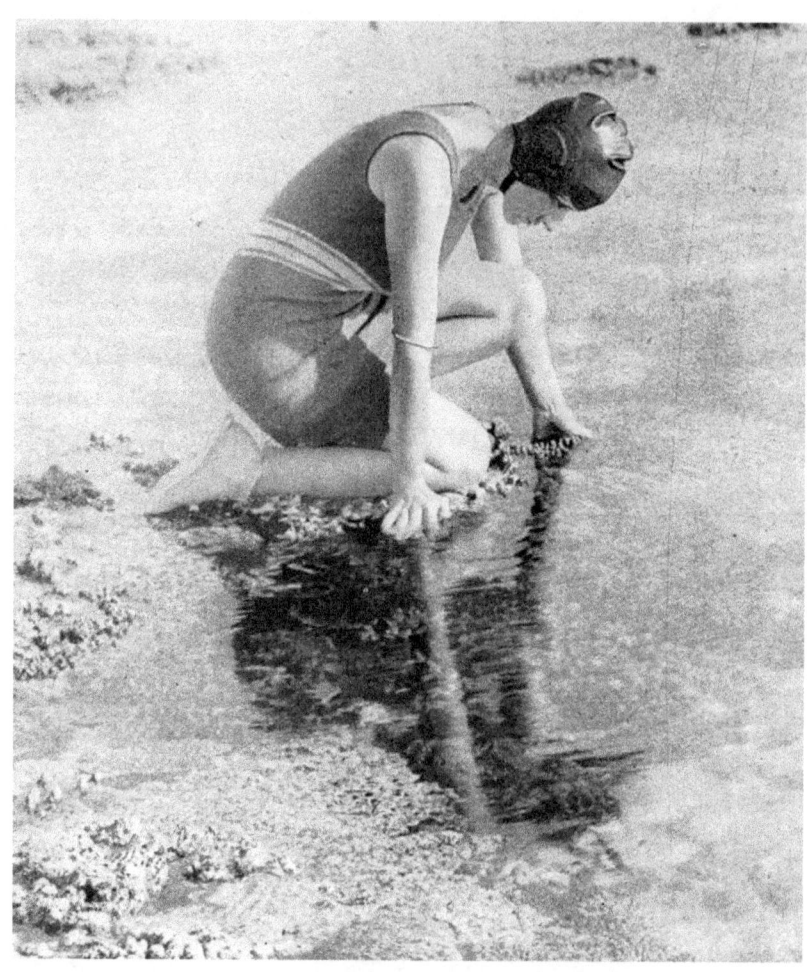

Fig. 1.1. The Great Barrier Reef, *Home*, May 1, 1928, 38.
Photograph by Anthony Musgrave.

CHAPTER 1

Coral Empire

In the 1920s, when the photograph in figure 1.1 was taken, oceans and the undersea were still, as they had been for centuries, the great enigma and frontier for poetic and technological exploration. Westerners imagined the deep sea and coral islands as wilderness external to human beings and separate from technology, culture, and daily life. And the underwater, whether at the edge of the sea or in the depths, represented an uncanny world of hidden animals and invisible mysteries. The sea confirmed already held beliefs based in the dualisms of human and animal, culture and nature, material and immaterial, and triggered questions about the boundaries of human and nonhuman life.

By the 1920s, the science of coral reefs had established once and for all that corals are not flowers, insects, or worms, as they were once thought. In the 1920s, the reef, once an object of bewilderment and fear, was reimagined through dreamy images of submarine glories and ranked as "unexcelled for beauty among all the spectacles of the universe."[1] In 1928, when the woman in figure 1.1 was alive, there was great anticipation of the aesthetic enrichment and pleasure that coral reefs would bring to peoples' lives. In 1928, for instance, the American marine zoologist William Beebe (1877–1962) urged each of his readers to witness firsthand the beauty of a coral reef. He wrote, "Don't die without having borrowed, stolen, purchased or made a helmet of sorts, to glimpse for yourself this . . . unsuspected realm of gorgeous life and color existing with us today on the self-same planet Earth."[2]

But, of course, it had not always been the case that coral reefs were conceived as beautiful objects of gorgeous form and color. They have enriched but also ruined human lives, enabled and also destroyed human endeavors. "Woe to the ship which in an ebbing tide, and with a strong wind, may be driven across some of these subterranean ridges" warned the *Queensland Times* in Brisbane, Australia, in 1875.[3] Captain James Cook (1728–1779) called the Great Barrier Reef of Australia an "insane Labyrinth" of coral. But Iain McCalman suggests how insane Cook himself was to sail the reef at night, a decision that saw his ship, the *Endeavour*, marooned for five weeks. That near disastrous decision, argues McCalman, was one of Cook's "profound environmental misunderstandings."[4] This legendary moment of human and coral contact, which is also part of the history of science, exploration, and British expansion, illustrates vividly how the relationship of corals with history and culture has always been important. Corals epitomize "vibrant matter," the type of nonhuman bodies that Jane Bennett characterizes as having the capacity to "impede or block the will and designs of humans but also to act as quasi agents or forces with trajectories, propensities, or tendencies of their own."[5] Corals have always been a physical and cultural force.

So strong a hold have corals exerted on the popular imagination that Stefan Helmreich refers to coral reefs as "figures," by which he means they are made by "creatures of fact and fiction that symbolize and embody social and scientific tensions, trends, and transformations."[6] Drawing on Donna Haraway's work in the history of science and human-animal relations, Helmreich's study of the figure of the coral reef discusses the extent to which this oceanic phenomenon is prefigured and written in our minds. Foremost among the cultural avenues for the reef becoming an imaginative force was the popular press of the nineteenth and early twentieth centuries, where the subject of the reef preoccupied news of science and navigation, where a great deal was written about the beauty of coral reefs in tropical waters, and where much in evidence was the appropriation of the coral reef as metaphor for imperial ambitions and empire building.

As single, separate, and minute beings, there has been little about corals with which people have been able to identify. Charles Darwin (1809–1882), for example, described how a coral reef is built by tiny, "apparently insignificant," gelatinous creatures.[7] Today, Eva Hayward argues that coral organisms are so tiny that we, as humans of a much bigger size, cannot map our bodies onto them, and as a result, it "makes identification a politics of erasure rather than empathy."[8] But as colonies, as collectives, corals lend themselves easily to human identification. That is because they are characterized as empire builders and colonial animals.[9] Humankind can easily identify with corals through militarized and anthropomorphic metaphors.

When Charles Darwin proposed the theory of subsidence to explain how volcanic islands sink, and also proposed the idea that corals develop upward at the edge of a sinking island, growing toward the light to become barrier reefs and atolls, he solved a scientific problem. But Darwin's theory of coral reef formation by tiny coral animals also became a suggestive metaphor for the ambitions of the British colonial project, and associated ambitions of Christian missionary life. Concerning the moral metaphor that developed around coral, Michelle Elleray explains how Darwin's theories acted as a social catalyst, and how in the mid-nineteenth century, "discussion of coral moves from scientific debate into a discourse on Victorian moral codes, especially in evangelical circles," where Christian families, for example, prepared their children for missionary work by likening them to the virtuous "insect" builders of the reef.[10] But Darwin's influence also shone a light on what was popularly called the "industry" and "work" of coral animals. These anthropocentric conceptualizations explain why corals came to be known as "the greatest builders of the world," as if to say they live their lives in the same way as human beings: purposively making a worthwhile life through the design and construction of colonies, settlements, and communities.[11]

By the end of the nineteenth century the British Empire boasted ownership of the world's greatest underwater coral empires. Newspapers, acting as ciphers for social and colonial perceptions, were keen to point out the extent of Britain's coral island possessions, and in 1917, a point in history that E. J. Hobsbawm argues marks the end of the British imperial project, one article claimed "there are over ten thousand islands in the British Empire."[12] It was 1718 when Britain acquired the Bahamas, and 1859 when Queensland, Australia, whose shores are girded by the Great Barrier Reef, became a colony of the British Empire. The colonial project harnessed the moral dimension of the figure of the coral reef as the product of industrious and spiritually dedicated workers, and combined it with the politics of an international ambition for social, military, and economic power. The imperial imaginary found in the figure of the coral reef a useful political image and a metaphorical space to assert the rightness and goodness of the empire's own colonizing practice of expansion. Acquiring and building colonies, especially in the tropics, seemed as organic for the British Empire as the process of reef building itself. And, in an age of positivist science and Enlightenment influence, a marine animal that also built toward the light embodied a useful social symbol for enlightened Europeans.

However, a coral reef was both a robust and a risky metaphor for serving imperialism. It was a shaky comparison because coral reefs and empires are built over time, and while both are capable of rapid expansion, both are also susceptible to decline and ruin. Moreover, as history has shown, there is nothing more ambiguous than corals. On one hand, they are an undeniable force of nature; on the other, they represent a confusing entity that seems to defy the boundaries between male and female, and animal, vegetable, and mineral. Stefan Helmreich makes the point that encounters between humans and corals have frequently been based in disorientation and misunderstanding.[13] And in 1972 Samuel M. Weber spelled out the nature of the epistemological confusion: "Animal, vegetable or mineral? The meanings of 'madrepore' are themselves strangely madreporic: animal, vegetable and mineral, living and dead, producer and product. Only its porosity seems beyond question: yet, here too, a certain confusion appears."[14]

Plant, stone, animal: the cultural history of corals shows they have been classified as all three. Flowers, insects, worms—corals have been confused with each one. Confusion over the true nature of corals came partly from science and partly from myths such as the myth of Medusa, who turned the plants of the sea to stone.[15] Coral's fluid, visual boundaries and resemblance to flowers and gardens once assigned it to the vegetable kingdom until science later revealed it as the work of colonizing animals. In 1936, the children's writer Frances Jenkins Olcott called them "flower animals" to acknowledge their hybrid, border-crossing nature.[16]

Possibly as a consequence of their often contradictory nature, coral reefs were models for a wide variety of social claims about the British Empire, In 1861, they were a demonstration of the correct structure of a colonial society in which "the broader the base, the loftier the apex."[17] In 1908, they were seen as mirrors of the human character, which, "like a coral reef, is made bit by bit."[18] They served as a cautionary tale for the potential chaos and randomness of expansionism, with some observers concluding that the British Empire "grew like a coral reef, without a plan."[19] They justified the significance of brotherhoods, guilds, and fraternities because "society has been built up like a coral atoll of innumerable fraternities—social, political and industrial."[20] And when, in 1929, there were mounting concerns across the empire about worker exploitation, coral reefs served as a threat and warning to anyone who would forget that "[the workers] build the reef, and the reef maintains them. Disaster to the reef means death to its inhabitants."[21]

The coral reef analogy continued to hold currency well into the early twentieth century. In 1905, coral polyps and their "work" offered a conceptual validation for the ongoing processes of colonization, for democracy, and for the greatness of the British Empire: "The Empire of Britain, like the coral reefs that guard so many of her island possessions, has been built up by no one man, but is the united product of the efforts of a countless multitude whose works alone abide. Statesman and diplomatist, sea-rover and soldier, historian and poet, administrator and the sturdy son of the soil, have all given their lives to the great work of empire-building."[22] However, conceiving of the British Empire as a reef built by politicians, military figures, artists, historians, farmers, and mariners betrayed it as a racially exclusive democracy. Excluded was the contribution of Indigenous labor in building "the reef." As colonies of exploitation, places such as the Bahamas and Australia took advantage of local labor and resources to help make Britain flourish, and colonial newspapers acted as ciphers for the systemic racism that often characterized coral island peoples as inferior. When the Gilbert and Ellice Islands (now the separate countries of Kiribati and Tuvalu) were annexed as crown colonies of the British Empire in 1915, a headline read, "More 'Lumps of Coral.'"[23] It was a figure of speech that belonged to a vocabulary of domination: "lumps of coral" portrayed the people and culture of the islands as an indiscriminate conglomerate of things rather than a social ecology of differentiated individuals. Not only islands of the Pacific but also islands of the Atlantic, particularly the Bahamas, were known in colonial circles as "little unfruitful bits of soil."[24] But what the islands of the Bahamas were praised for was their value as ports to hold the empire together through sea power.

James Cook's journals planted in peoples' minds the idea that coral reefs are fearful, disorienting, monstrous human snares. But that image changed greatly in the late nineteenth century when the impact of empirical science and the pleasures of observing the natural world from life turned attention to the reef as a living entity. This shift in perspective and interest brought increasing attention to what reefs looked like from under the water as well as from the surface. In the early twentieth century, people became more interested in the natural environments of coral reefs without the encumbering symbolisms of empire and colonization. The colonial lifestyle of corals was of interest in and of itself without allusion to the genesis of human colonies. People wanted to know how coral animals breathed, what gave them form and color, their position in the animal world, and how atolls were made. As anthropomorphism gave way to increasing interest in the autonomy of animals and plants, coral reef environments such as the Great Barrier Reef were described in nonhuman terms as "the greatest animal structure on this planet."[25] Sydney Elliot Napier (1870–1940), a poet and journalist who explored the Great Barrier Reef in 1928 with a party of people that included the woman in figure 1.1, said it was inconceivable that people had ever referred to corals as lowly creatures and had looked upon corals as "so small, so helpless, so weak and seemingly 'contemptible'" that they had also found it difficult to believe that coral animals could build such complex and extensive habitats.[26] Instead, Napier found it sublime and decentering to think about the deep time associated with reef building and the ancient nature of coral formations: it is, he said, "a miracle; a thing to strike us dumb."[27]

The beauties and dangers of coral reefs and the relative accessibility of the underwater zone surrounding reefs presented a unique opportunity for cinematic and photographic representations. A thoroughly modern relationship developed between photography and filmmaking, coral reefs, and tropical water, one based in the common ingredient of light and the material quality of transparency. It became apparent that there was a magical correspondence in the way the natural phenomenon of corals and the technological processes of photographs both required light for photochemical reactions. It was a revelation that corals, as well as photographs, needed light to bring them to life and enable development.[28]

Moreover, photography and the tropics seemed like a perfect marriage because tropical water in the right conditions seems almost as transparent and clear as air. The more transparent the water, the more visible the creatures and plants of the underwater, and the more successful the fixing of photographic images, moving and still. With the environment of the tropics readily engaging minds about the limits and potential for vision through water, the tropics became a space of the imagination similar to the technological spaces of modern entertainment. Like the illuminations of magic lantern slides and cinema screens, the water of the tropics was a site of projections, mirroring, illusions, and luminous shadows. The way the medium of water offered a lens to apparitions seemingly projected below in the underwater was similar to watching cinematic projections and recalling dreams.

The reef was an outer world on which the whole gamut of human emotions could be projected. It served as a source of social and political metaphor and as an inspiration for poetic vision, and it acted as an agent driving changes in technologies, shaping cinematography and photography, luring explorers and travelers, and forging new visual experiences for spectators of images. Photographers and cinematographers saw an opportunity to generate on film a coral orientalism that would appeal to general audiences, the science community, and maybe also artistic modernism. But to achieve this they had to be explorers as well as image hunters. Felix Driver points out that in the twentieth century, a distinction between the "adventurous explorer and the scientific traveller" came about through the modern explorer's links with cinema, mass media, sensation, the exotic, and "the inexorable advance of technological and commercial modernity across the globe."[29]

Coral reef environments opened the mind to the "floor of the sea" or the "bed of the ocean" in a positive way. A fixation on the underwater of reefs emerged coterminously with cinematic society when interest in the natural world intensified with the advent of the moving image, turning the coral reef and the underwater into objects of a voyeuristic gaze. In the spatialization of the sea, the deep sea has been conceived as a sublime wilderness, while the shallower waters surrounding coral reefs promised a wilderness that was relatively benign because it is where water, air, land, plants, rocks, animals, and humans meet. Unlike the deep sea, which hides the abyss, the shallow waters of tropical fringing reefs and atolls were welcome for their vibrant features. That is not to say they did not have their own real and imaginary dangers in the form of hidden and alien animals. But the limestone rock that is the reef also promised safety and escape, whereas the deep ocean offered only disorientation and terror. Terrestrial space has mountains, jungles, and polar continents that stand out in the landscape and beckon to explorers. These are defined and definable features that anchor the imagination to singular environments. The deep sea is featureless, but a coral reef, as Jonathan Crylen points out, is a "static spatial landmark" in an otherwise undifferentiated environment.[30] And scientists also agree that the pelagic zone of the open sea is an ambiguous space where the "biomass density in open water is minute compared to reefs."[31] Coral reefs give explorers and photographers a tangible, featureful space to focus their minds.

On a clear, still day, at the edge of a reef, the eye can gaze through the water to the spectacle below because light filters through shallow water differently than it does in open sea. Shallow, transparent tropical waters are reassuring because they make a promise that the eye will be able to make sense of every object. Transparency and certainty are the qualities that made tropical water legendary. And color. In coral reef environments, the visual ecology pulsates with colors, patterns, and textures. The spectral reflections of yellow and blue reef fish, for example, make for some of the natural world's most flamboyant displays. This much was pointed out by E. M. Stephenson, a midtwentieth-century biologist who referred to the coral reef as a stage, not in the sense of an inert background but rather as a dynamic system of species relations. Describing the lustrous shapes and colors of underwater animals and plants, she explained to the general reader how "coral fish are of the most varied and brilliant colors imaginable, as indeed they must be if they are to blend with the rainbow stage on which their lives are set."[32]

Plate 1, an underwater photograph of a healthy reef at the Australian Great Barrier taken by a scientist in 2017, brings the reader closer to the idea of a coral reef as a rainbow stage on which the lives of underwater animals and plants are acted out. This is the vision, the fantasy, and the aesthetic that lured many an early photographer and cinematographer to the tropics in the early twentieth century. They went there in hope of immersing themselves in the underwater to capture the spectacle eye to eye. Any number of dramas might present themselves on the stage of the coral reef but, as E. M. Stephenson explained by citing the research of the English artist and author, Robert Gibbings (1889–1958), the human viewer had to get close because a coral reef is a camouflage phenomenon.[33] In the theory of camouflage put forward by Charles Darwin, colorings and markings on animals are an adaptation for survival that enable organisms to hide from predators through the process of crypsis (blending) and mimicry (imitation). Both processes serve the function of concealment and deception.[34]

For every fish that is visible in plate 1 there are hundreds more that go un-seen among the labyrinthine forms of coral that seemingly blossom and sprout as if "flower beds under the sea."[35]

Viewed from above the surface of the water, from the edge of a boat or the edge of a rock pool—in fact, the viewing point taken by the woman in figure 1.1—a reef below will look foreshortened and abstracted. The colors of the reef will look different above water and below. When coral reefs are submerged, and covered with seawater, and the viewer looks down, the colors appear muted—according to Robert Gibbings who described the optical experience of viewing reefs in Bermuda in the 1930s. Well before the popularity and widespread use of self-contained underwater breath-ing apparatus (scuba) for diving emerged, Gibbings was a firm believer that to experience a coral reef required embodied observation. To immerse the eyes and body in a tropical coral reef meant getting below the surface. Using a helmet with air hose, Gibbings submerged at Bermuda to make drawings of the undersea, and in a later book, *Blue Angels and Whales* (1938), described what it was like to get on "closer terms with the fish, and to meet them on their own level."[36]

The colors, Gibbings said, when the reef is viewed from the surface of the water, "take on the mellowness of an old master rather than the cru-dity of a new one and the fish are no more obvious than a mallard among reeds, or a butterfly at rest in a garden."[37] The reef seen from above does not reveal its particularities or details. The human observer may think he or she will see everything revealed through clear water, but instead things seem to vanish. Even the brightest-colored creatures will match the sur-rounding reef and "disappear" in plain sight when the light hits their dorsal side, or evade the eye of the observer when the broken patterns on their skin match the tones of other beings around them.[38] The colors and pat-terns of coral reef environments are so innovative, and so remarkable, they have also given many human observers, such as the philosopher Alphonso Lingis, pause to consider a less functionalist explanation than camouflage: perhaps, suggests Lingis, all the "frivolity" is for the delight of the nonhu-man world.[39]

An interest in the wild underwater, and in coral reefs and their repre-sentation in art and science, emerged in European history in the after-math of the First World War. Is this timing significant? Was it a need to find reenchantment in the world through beauty and the transcendent spectacle of nature that encouraged people to look differently at coral reefs and the underwater? Was it the war itself, and submarine warfare in particular, that led to the spatial zone of the underwater being imagined anew? Is there a connection between the militarization of the underwater in the First World War by submarines and the emerging preoccupation with the undersea and penetration by cameras? The submarine camera and the military submarine—both were technologies for spying. For early twentieth-century photographers, the idea of spying in the sea was coupled with the fantasy of seeing the underwater for the first time: "There is little to be seen on the surface of the globe these days," wrote Frank Hurley in 1926 during one of many adventures to the coral reefs of the Great Barrier, "but the sea-floor opens up limitless avenues to our inquisitiveness."[40]

What can be said by way of addressing the questions posed above is that the military colonization of the undersea in the First World War, coupled with the first forays into aerial warfare, expanded every citizen's scale of vision. It became more commonplace to imagine the human body's position in a vertical space that was no longer simply terrestrial but included the sky and the undersea as well. In the aftermath of the war, in 1928, the same year that the woman in figure 1.1 gazed into the depths of a coral pool at the Great Barrier Reef, the French surrealist artist André Breton captured the spirit of the times when he wrote about the freedom as well as the disorientation embodied in imagining the scale of a planet not just as a terrestrial world but also as a place of wonder that encompasses "the Marvels of the earth a hundred feet high, the Marvels of the sea a hundred feet deep."[41]

In 1928, as interest in the floor of the sea had intensified, the underwater had not yet become a tangible part of photographic and filmic experience, nor become a visible part of the photographic and filmic record. But that was about to change, and Breton and surrealism were involved in the change. In various ways, through art making and publication, they worked with imagery that drew attention to the mythical, visual, and imaginative spectacle of the underwater and of coral reefs. Through mass media in the early twentieth century, the "coral empire" that was shaped by social and physical interactions between colonialists and coral colonies turned into an empire of coral images that were released in the world through the expanding media world of magazines, newspapers, advertisements, cinema, and postcards.

Roy Porter explains the term "the empire of images" as a cultural phenomenon that happened in the twentieth century with the emergence of modern consumer society. He refers to it as revolutionizing twentieth-century consciousness.[42] Through the circulation of images in popular culture, especially after 1928, Western audiences, scientists, and artists came to know "nature," "the underwater," and "the coral reef." But, as commonly happens in the empire of images, photographs are taken from their original context and placed in new contexts, and when this happens they acquire new meanings.

For centuries, explorers and artists had represented the natural world through drawings and engravings. Well before photography, for example, the naturalist Alexander von Humboldt (1769–1859) understood very well that "the world likes to *see*."[43] Humboldt felt the urge to bring the natural world to the public through visual images that, before the use of cameras, were drawings that were also turned into etchings. But natural history changed in the early twentieth century through the scale of photographic distribution. The industry of mass communication that accelerated in the early twentieth century created a volume of images for distribution that was unprecedented.

Through the growth of the empire of images, the Western world was able to experience "the marvelous ocean floor" from home or from the comfort of the cinema screen—according to a newspaper article about the undersea explorer and photographer J. E. Williamson, who by 1925 was world famous for photographing the Bahamas underwater from a submersible.[44] The shallow waters around Williamson's home, as distinct from the deep waters of the middle ocean, made his enterprise possible. In shallow and clear water, he was able to sit on the seabed, in an air-filled chamber, and photograph the marine world as it swam by. Over the next twenty years, Williamson's still photographs would appear in newspapers and periodicals in Australia, the United States, Canada, and Britain. The spectacle he filmed and photographed in the 1920s was so new, it hardly mattered that the reefs were seen in the gray tones of black-and-white photography in the years before film and camera technologies would routinely reproduce coral reefs in color. Through Williamson's work, the coral imaginary, like the underwater imaginary, was shaped and packaged through photographs and films that were products of a visual modernity ruled by the scopic regime, or the drive to see all. Through Williamson's work the figure of the coral reef became likened to a morphic dream—a hallucination of riotous shades and forms—suited to modern spectacle and what Siegfried Kracauer refers to as "the dreamlike subjectivity of the cinema spectator."[45]

Something we take for granted today about underwater photography is the ability and desirability of an embodied sense of diving and photographing in the underwater. But in the 1920s, photographing the underwater was an almost insurmountable problem. Contrivances had to be used that surrounded the photographer with air, devices such as submersibles and aquariums. Using a submersible, Williamson and his brother George had assisted with underwater camera work for the world's first underwater motion picture produced by Universal Pictures, a 1916 adaption directed by Stuart Paton of Jules Verne's novel *Twenty Thousand Leagues under the Sea* (1870). Using a submersible, Williamson had helped the American Museum of Natural History and the Field Museum collect animals and marine specimens for their collections and for taxidermy. He published still photographs of the underwater in the popular press.

In 1937, André Breton found among the stock images of the *New York Times* a photograph of an underwater Bahamian coral reef taken by J. E. Williamson. Shot in the Bahamas but distributed by the *New York Times* to international photo agencies, including one in Paris, the photograph was a symptom of the cultural expansion of America into the everyday life of the Bahamas, and the expansion of America into mass media, as well as a sign of the expansion of the empire of images throughout the Western world. Breton published the image in *Mad Love*. In the first chapter of that book he compared the properties of coral and crystal and developed a theory of aesthetics he called "convulsive beauty." He wrote, "Life, in its constant formation and destruction, seems to me never better framed for the human eye than between the hedges of blue titmouses of aragonite and 'The treasure bridge of the Australian "Great Barrier."'"[46]

The image had once belonged to a specific time and place, but reproduced in the poetic context of *Mad Love*, it was set loose from any realist anchors. The surrealists, who did not follow the main intellectual trajectory of modernism into abstraction but were attracted by figuration, dream images, and hallucinations, were also attracted to the rococo nature of the underwater organic.[47] The visual spontaneity of coral reefs suggests life in its most spontaneous form, which appealed to a movement grounded conceptually in chance and automatism, and attracted to the uncanny mimicries of nature. Because coral reefs grow underwater, they appealed to an artistic movement committed to the symbolic realm of the unconscious. The materialization of the invisible space of the underwater was, to surrealism, the embrace of the principle and aesthetic of making strange, an effect of disorientation made even more emphatic by the often ghostly, gothic shapes of coral reef formations. As explorers of the mind, surrealists latched onto the metaphoric potential of the figure of the underwater coral reef. By examining, now, the case of Breton's acquisition of J. E. Williamson's photograph of a coral reef underwater at the Bahamas, which he renamed "the Australian Great Barrier," we will arrive at a point of understanding more about the significance of coral reefs and the underwater to modern visual culture, and more about the life and work of J. E. Williamson.

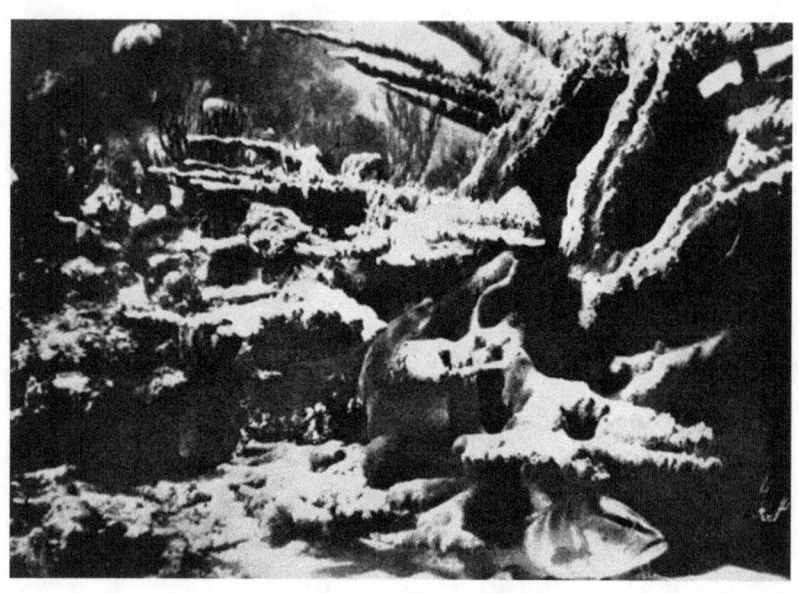

Fig. 2.1. "Le Pont de Trésors de la 'Grande Barrière' Australienne."
In André Breton, *L'Amour fou*, 1937, 27. Photograph by N.-Y.-T.

CHAPTER 2

Mad Love

"Human explorers" is the term André Breton used to characterize the artists who were part of the movement he founded in 1924. Surrealism was an artistic and political rebellion arising from disillusionment with the social world in the wake of the First World War. From that time on, and under Breton's influence, surrealism was interested in encounters with the irrational, the unconscious mind, and imaginative states realized through dreams. It was in the first *Manifesto of Surrealism* (1924) that Breton referred to himself and progressive thinkers as "human explorers" whose work was to investigate the imaginative possibilities of the unconscious in reaction to "the pretense of civilization and progress."[1]

To liberate the mind and body, as well as thought and expression, through chance and automatic modes of thought and action, and to free the self from the constraints of rationalism was surrealism's manifesto. Which is why the sea, a shifting, unfathomable force of nature, was totemic for the movement and why the underwater was revered as a symbol of the unconscious. Allan Sekula (1951–2013) named surrealism "the last aesthetic movement to claim the sea with any seriousness."[2] And his point reverberates through a survey of the most evocative works of art and literature in history, among them Max Ernst's *Forêt* (1927), René Magritte's *Collective Invention* (1934), and René Clair's *Entr'acte* (1924) a short, silent film in which a dancer, viewed from below in slow motion, appears weightless as her petticoats undulate like a jellyfish in water. There are the examples of Man Ray's use of underwater films by the biologist Jean Painlevé (1902–1989) for *L'Étoile de mer* (1928) and the inclusion by Georges Bataille of Painlevé's pictures of crustaceans in the journal *Documents*. Moreover, it was in Breton's essay "Surrealism and Painting" (1928) that he wrote about "the Marvels of the sea a hundred feet deep," and how only the "wild eye" freed from habit can be fully receptive to the magical sensations of the outer limits of the world.[3]

Sometime before 1937, André Breton obtained an underwater photograph from Wide World Photos, the photo agency then owned by the *New York Times*.[4] Underwater photographs were only just beginning to appear in natural history publications. They were relatively rare, but a keen public waited to view them.[5] In 1937, Breton published the news photograph in the pages of *Mad Love* (see fig. 2.1).

How did Breton, and for that matter, the *New York Times*, obtain the image? In 1937, very few underwater photographs of the Great Barrier Reef existed, and even fewer were published. Australians themselves had not shown great interest in the Great Barrier Reef from the perspective of the underwater because Australian settler identity, history, and national outlook were oriented toward the myths and logic of the terrestrial world. How the photograph first came to Breton's attention is also unclear. A fellow surrealist might have passed it on, as happened throughout the history of the movement when artists found surprising press photographs.[6] But one thing is certain: at some point Breton obtained a photographic print from the Paris office of the *New York Times*. This much is clear because the print remained in Breton's collection of books, manuscripts, photographs, and art at 42 rue Fontaine in Paris until his death in 1966, after which the Association Atelier André Breton digitized the underwater photograph, along with the artist's prints and letters.[7] On the verso side of the print is the stamp of the Paris office.

The photographic print in Breton's collection reveals the image in its original form at the time Breton aquired it from the *New York Times* and before he cropped it for publication in *Mad Love* (fig. 2.2). Both the front and the back of the photo agency print are marked with Breton's handwriting. On the front, below a line drawn horizontally across the image, a note provides instructions on how to crop the photograph. In the final reproduction seen in figure 2.1, the lower section is missing. Cropped out for publication in *Mad Love* is the bottom of the image showing what looks

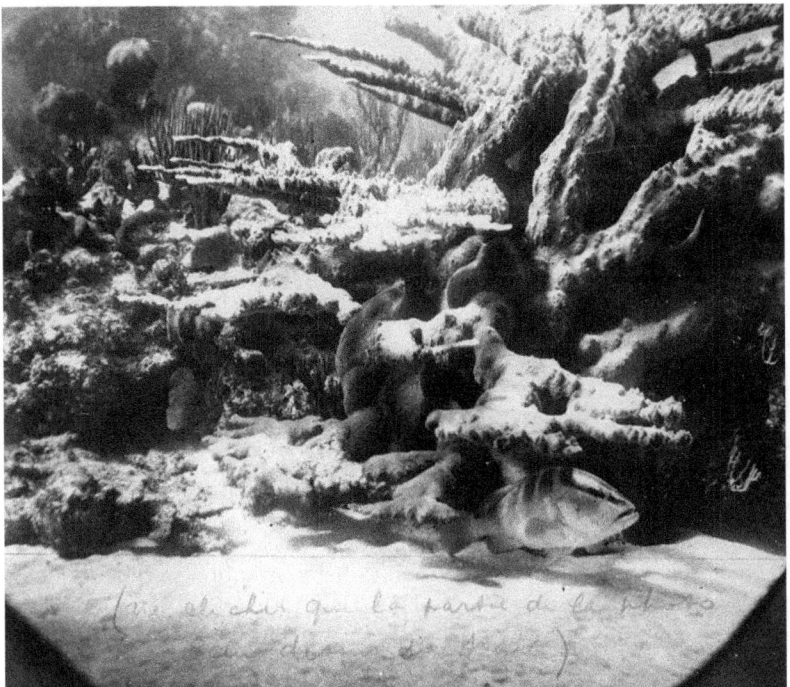

Fig. 2.2. Front of Wide World Photos photographic print acquired by
André Breton, Photo: Association Atelier André Breton.

like the curved edge of a window through which we see a distorted fish-
eye perspective of a white ocean floor.

A quick check of the digitized *New York Times* reveals the context in
which the underwater photograph that Breton acquired was published.
It featured in an article written in 1929 titled "The Coral World beneath
the Waves That the Williamsons Invaded."[8] The photograph that Breton
later labeled "The Treasure Bridge of the Australian Great Barrier" was
actually taken underwater at the Bahamas during an expedition there in
1929 organized by the Field Museum in Chicago in conjunction with the
explorer, photographer, and filmmaker John Ernest Williamson to collect
corals, sharks, and other fish for the museum's newly planned diorama
scene of a Bahamian coral reef.[9]

The newspaper story told how J. E. Williamson was an expatriate American living in Nassau on New Providence Island in the Bahamas with his
wife and daughter. When Williamson teamed up with the Field Museum,

he had organized a contract with the *New York Times*—underwater photographs, being rare, were a valuable commodity, and these would be never-before-seen views. Virginia Pope, the journalist who wrote the article in 1929, set the scene: "In the depths of the ocean off the Bahama Islands lies a fantastic world. Few men have seen it. . . . The coral trees and the fishes, with their moods and their manners, have few secrets left now that the Williamson trio—Mr. and Mrs. J. E. Williamson and baby Sylvia— have penetrated into the hidden grottoes where they pass their days and nights in under-water seclusion, looking through the cyclopean eye of their submarine studio."[10]

The story featured the methods that J. E. Williamson used to obtain underwater photographs and described early collaborations with his father, captain Charles Williamson, and brother, George Maurice Williamson. The captain had designed a submersible that could be operated underwater to enable the salvage of shipwrecks. Later, though, J. E., with the help of his brother, had adapted their father's invention so they could explore and photograph the sea floor. They added a spherical steel appendage to the submersible and called it a "photosphere." It was an air-filled chamber with a round glass window—described in Pope's article as a "cyclopean eye"— through which they filmed and photographed, and it was pulled along underwater by a barge called *Jules Verne*.

As the naming of the barge suggests, J. E. Williamson was a Jules Verne enthusiast. In 1929, when the *New York Times* article was published, he and George were already famous for directing underwater scenes for 20,000 *Leagues under the Sea* (1916), the Universal Pictures adaptation of Verne's book. The film was directed by Stuart Paton, and many scenes were shot in the Bahamas. It was released in France after the First World War, and in every likelihood it was seen by André Breton and the surrealist group. Margaret Drabble argues that Verne's book *Twenty Thousand Leagues under the Sea* shaped the French avant-garde.[11] Pamela Kort observes that it was Jules Verne who motivated the surrealist's fascination with the underwater realm. She claims it was *Twenty Thousand Leagues under the Sea* that inspired Max Ernst's paintings of forests and that *Forêt* was a response to Verne's passage about the resemblance of corals to petrified trees. This explains, writes Kort, why viewers of Ernst's works feel as if they are "standing on the ocean's floor."[12] The combination of circumstances, including the revolutionary nature of Paton's film coupled with the French passion for Jules Verne, strongly suggests that the news photograph that Breton obtained from the *New York Times* was not his first exposure to the work of J. E. Williamson, although this fact he may not have registered.

At the start of Paton's film, a title card reads, "The submarine scenes in this production were made possible by the use of the Williamson inventions, and were directed under the personal supervision of the Williamson brothers, who alone have solved the secret of under-the-ocean photography."[13] Solving the secret of underwater photography was in fact an aspiration straight from Verne's novel. J. E. Williamson had so thoroughly embodied the perspectives of characters in Verne's story that his approach to the underwater is difficult to distinguish from attitudes and personas that define *Twenty Thousand Leagues under the Sea*. The concept of standing before a glass window looking out through the ocean and taking photographs mimics the scene in the book in which Captain Nemo, training a camera on the vision unfolding before the Nautilus's window, comments to Aronnax that "nothing could be easier than taking a photograph of this underwater region."[14] Natascha Adamowsky explains in her account of the symbolism of the underwater window in *Twenty Thousand Leagues under the Sea* that in both cases the window acted as "a screen, a window space, a frame and a showcase" onto the alien sea.[15] The window was an interface to the aquatic realm beyond and epitomized an aquarium perspective.

Verne himself is known to have been inspired by visits to aquariums. They shaped how he visualized the underwater realm in *Twenty Thousand Leagues under the Sea*.[16] In 1870, it was routine for people to photograph marine life in domestic aquariums and call them "underwater" photographs. Aquarium photography was still practiced in the 1920s, including in surrealist circles. Jean Painlevé, for example, filmed underwater marine life in a tank but made the scenes look like they were taken underwater.[17] Although Painlevé had devised a waterproof camera and breathing mechanism to film sea creatures in the Bay of Arcachon, he also used an aquarium for filming sea creatures up close, including a seahorse giving birth—films that have been described by Lauren E. Fretz as "part science, part poetry, part adventure."[18]

But it was in 1937, the same year *Mad Love* was published, that Painlevé truly excited the popular imagination with respect to the underwater. At the Paris World's Fair, he and Yves Le Prieur staged *l'Aquarium Humain*, a demonstration of underwater diving and swimming performed in a glass tank. Painlevé wore extraordinary rubber foot fins designed by Louis de Corlieu and breathed compressed air in a bottle invented by Le Prieur.[19] Like a cyborg, he postured as part human, part machine, and part animal.[20] Humans had finally become amphibious, if only in an artificial way, since their return to nature was through technology. Or perhaps it was not a sign of a return to nature so much as conquering nature. As Jonathan C. Crylen has recently argued in relation to Jacques-Yves Cousteau (1910–1997)—the next generation of French diver—he had "beaten otherness at its own game: inhabiting the fluid space of the sea more dexterously than could other bodies and thus convert[ing] it from a space of difference to one of continuity with the industrialized West."[21]

Through the sides of the glass tank, the public could see how the underwater challenged the known world, decentered the human subject, and exposed how modernity had become synonymous with the colonization of the sea through technology. In Painlevé's public display of these ideas, and in making the underwater a place of scientific study and popular entertainment, he paralleled the achievements of J. E. Williamson. Eventually, in the 1950s, both men would be recognized as pioneers in books detailing the history of European underwater exploration and photography.[22]

Technology-aided underwater diving and underwater photography were among the most talked about and intriguing developments of modernity. They enabled a new viewpoint on the world. André Breton wanted to be part of history in the making, and Williamson's photograph was the conduit. He published the image in *Mad Love* as an anonymous press photograph, yet Williamson's name was typed on the back of the print he obtained from Wide World Photos (fig. 2.3). Also on the back of the print is a caption that reads, "Expedition oceanographique J. E. Williamson dans les profondeurs de la mer flore et faune aquatiques, coraux géants, poissons, plantes marines scaphandriers, etc. . . ." ["J. E. Williamson oceanography expedition in deep sea aquatic flora and fauna, giant corals, fish, marine plants divers etc. . . ."]. But these words were crossed out as if irrelevant, superfluous, and unimportant. Written directly above the crossed-out caption, in Breton's handwriting, are the words "L'amour Fou," and on the opposite side of the back of the print back is more of Breton's writing: "Pl. 3.— Le pont de trésors de la 'grande barrière' australienne" (The treasure bridge of the Australian Great Barrier). This sentence was the basis for the final title in the first French edition of *L'Amour fou*.

The question arises of why Breton cropped out the bottom of the *New*

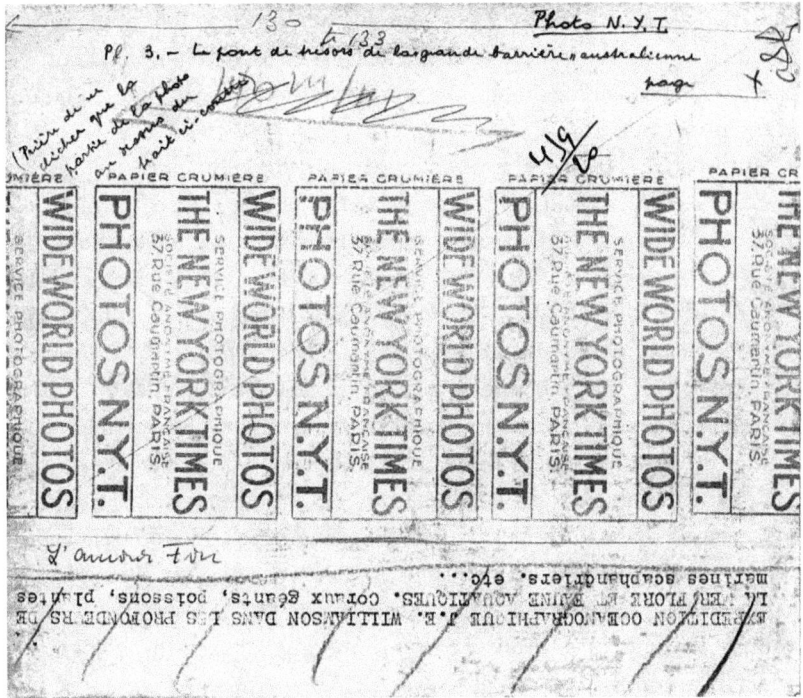

Fig. 2.3. Back of Wide World Photos photographic print acquired by André Breton, With title: "Pl.3.- Le pont de trésors de la 'grande barrière' australienne." Photo: Association Atelier André Breton.

York Times image showing the curved frame of the photosphere's round window. One answer might be that the curve, being a sign of the intermediary of technology in the production of the image, interfered with the sensation of an unmediated visual experience of the underwater realm and, moreover, that it interfered with the suggestion that the photograph was taken by a diver immersed in the underwater. The notion of an underwater diver was important to surrealism because a diver, like an artist, symbolized an explorer of the unconscious. To dive underwater was a sign of defiance and freedom from the rationality of daily life. It explains why Salvador Dalí, at the opening of the International Surrealist Exhibition in London in 1936, delivered a lecture in a deep-sea diving suit to broadcast the surrealist project, which was to "dive into the very depths of the human soul."[23] The imagined perspective of a diver-photographer immersed in the underwater inferred by the cropped photograph in *Mad*

Love increased the image's power. Breton cropped the bottom of the image because the original indicated something less exciting, even inauthentic: a photographer positioned behind a window rather than submerged bodily in the undersea.

The spatial clarity of the news agency photographic print (see figure 2.2) is the direct result of the scene being photographed through air. Taken from inside Williamson's photosphere as it was towed along the sea floor by the *Jules Verne*, with its passengers inside looking out at a marine world illuminated by tropical sun shining through shallow, Bahamian seas, and possibly aided by strong lights attached to the exterior of the photosphere, the photograph barely intimates that the medium outside was water. Instead, marine forms stand out in individual splendor. The sight is mediated by the photosphere's window, and the effect is similar to an aquarium perspective: vivid but strangely remote and exoticizing. The aquarium effect is described by Edward Eigen in relation to Louis Fabre-Domergue (1861–1940), an early photographer of aquarium life, as a photographic field relying on a "transparent frame" and "a picture window."[24] As Eigen argues in relation to Fabre-Domergue, aquarium photographs "visually annul the milieu" of seawater.[25] The point about enhanced visibility can be seen in figure 2.4—a film still from Williamson's documentary *Under the Sea* (1929). The vision we see is determined by the transparency of the photosphere's glass window, by the camera standing in air, and by the absence of silt normally stirred up by a photographer standing on the ocean floor.

Similarly, the news agency print that Breton acquired is clear and legible, the shades of gray subtle, and the depth fuller compared to the reproduction it became in *Mad Love*. On the printed page, contrasts of dark and light are abrupt. But in the uncropped news photograph, every mysterious flicker of light, every subtle variation in tone and pattern, and every eerie shape of coral reef flora and fauna is visible. The news agency photograph evokes a sense of peering into the depths of being and, in surrealist language, experiencing "the oceanic feeling" of subjectivity unconstrained. It gives material form to the passage in the first *Manifesto of Surrealism*, in which Breton observed how the "depths of our mind," like the underwater, contains unconscious forces, while the surface of the water symbolizes conscious, rational life.[26]

But once transferred to the printed book, the strange forms in Williamson's image lost clear distinction. What the image gained, however, was

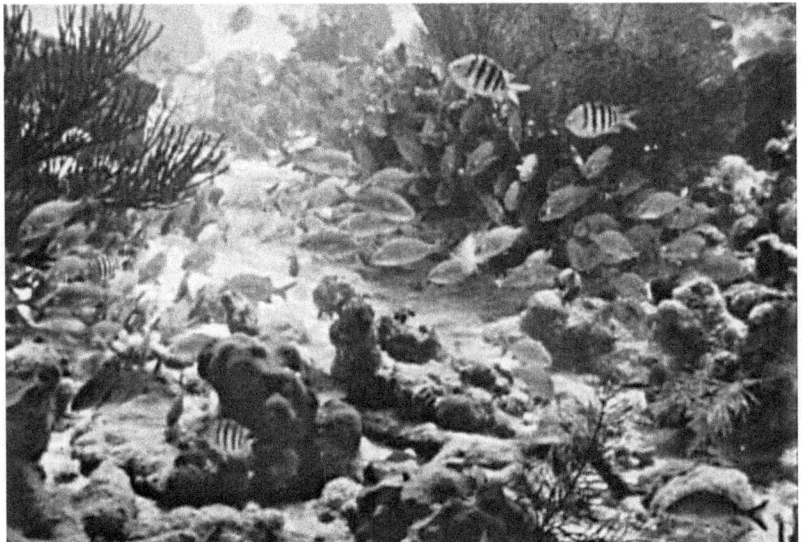

Fig. 2.4. Fish and corals. Film still. From J. E. Williamson, *Under the Sea*, 1929.
Courtesy of the Field Museum Archives.

a better sense of the disorientations of optics in the underwater. Recently Margaret Cohen explained how the indistinctness of the reproduction of Williamson's photograph in *Mad Love* is more typical of the material qualities of the underwater, the properties of which, she writes, "spectacularly defy the reality of terrestrial perception, along with the system of linear perspective."[27] She argues that the imprecise, hazy reproduction would have appealed to Breton and to "surrealism's search for imaginative expressions going beyond habitual terrestrial experience and intimating a higher surreality."[28] It is a visual effect, Cohen explains, that resonates with surrealism's notion of the *informe* as theorized by Georges Bataille as a kind of immersion in the world through the dissolution of bodily boundaries and the organizing principles of linear perspective. The relative value of underwater optics was a matter of debate in the early twentieth century. Alejandro Martínez explains how one side of the debate considered the aquatic medium an obstacle to comprehending submerged objects and believed in the necessity for new technologies to overcome poor visualization caused by the density of the materiality of water, yet the other side believed the form of the sea "was an inherent and innate property of underwater photography."[29] The conditions of viewing underwater are affected by suspended particles that prevent the kind of expansive depth of field and recession of perspectival lines that viewers had come to expect of conventional representations of landscapes seen through air. When explaining the differences between air optics and underwater optics, Trevor Norton observes that with underwater photography, "you might as well take photographs in lentil soup," especially when the camera picks out every speck of plankton and silt.[30] The point, then, is that the hazy reproduction in *Mad Love* has an authenticity about it that is connected to the actuality of underwater optics as it relates to human vision and to photography.

Breton's acquisition of Williamson's photograph is an example of the increasing internationalization of news via news agencies in the early twentieth century. By taking the image from the context of American journalism and recontextualizing it in *Mad Love*, Breton brought into play the surrealist practice of appropriating ready-made photographs from popular culture for surrealist ends.[31] But Breton's appropriation of Williamson's image also reveals how photographic images, when cut loose from their original context, will float in a virtual cultural space ready to attach to new contexts and signify new and different realities. By the stroke of a pen, the geographies of the Bahamas and the Great Barrier Reef were switched. It goes to prove Susan Sontag's point that when a photograph is freed from its historical and social moorings, it "drifts away into a soft abstract pastness, open to any kind of reading (or matching to other photographs)."[32] Yet what the surrealists gained by decontextualizing images from newspapers, magazines, and press agency stocks was a vibrant, endless source of the marvelous in the form of new relations between word and image.

The context of *Mad Love* offers clear answers for why Breton was attracted to Williamson's image. At a time before general audiences had developed a visual vocabulary for looking at the underwater realm, Williamson's films and photographs ushered in a new paradigm of seeing. André Breton was excited about the once invisible world beneath the sea coming to light. Moreover, Williamson's image of the underwater had a strong connection with surrealist politics and aesthetics because the vision it offered was like an otherworldly dream-space. The photograph was a significant find for Breton because it embodied the cornerstones of surrealist aesthetics—dreams, the unconscious, and the uncanny. It offered a vision that was opposite to the modern, metropolitan environments of Paris. If the experience of Paris was "geometric," the scene of "the Australian Great Barrier" underwater was an extravaganza of organic and natural wilderness of the type identified by Gavin Parkinson as "antithetical to the habits, customs, restrictions, and laws that characterized modern Western society."[33] The coral formations in Williamson's photograph that Breton published in Mad Love seem untamed and random. They invoke the suggestion of primordial mysteries surrounding the evolution of life on earth from the ocean, since it was commonly imagined that sea animals belonging to past ages would be found in the depths.[34] The image suggests not only regression but also the idea of unbounded experience. For Breton, who saw himself at the frontier of art, here was an image that also suggested the frontier of physical and psychological worlds.

But the link with the underwater and the feminine is also important because Mad Love is a tribute to Jacqueline Lamba, Breton's wife, whom he met in 1934 when she was a naked underwater dancer in a Montmartre music hall.[35] The naked female body represented in water, especially water suggestive of the sea, evokes, as Eldon N. Van Liere explains, "primal thoughts of birth and death, and for man both women and the sea touched his deepest emotions—his hopes for survival in the face of the unknown and his fears."[36] These were ideas explored later by Salvador Dalí in the Dream of Venus, Dalí's exhibit for the New York World's Fair of 1939. There he created a coral grotto that the public entered to find Venus asleep and reclining on a couch next to a deep water tank in which women were swimming. Venus, the goddess of love, was born of the sea. Dalí, who said he had memories of the intrauterine phase of his life, wanted to represent an unfolding dream imagined by the sleeping Venus.[37]

Mad Love also offers insight into the surrealist's attraction to the curious and paradoxical forms of corals. In Breton's own words, the intriguing dimension of corals lies in the way "the inanimate is so close to the animate that the imagination is free to play infinitely with these apparently mineral forms."[38] Corals appear to the eye as simultaneously animal, vegetable, and mineral, and Breton described the effect as an intense aesthetic experience of the "convulsive beauty" of marvelous things, "like the feeling of a feathery wind brushing across my temples to produce a real shiver."[39] Rosalind Krauss proposed that Breton's attraction to the mimicry of corals and their likeness to flowers, plants, and rocks stemmed from surrealism's interest in the creativity of nature, and in nature as producer of images and representations. Krauss concludes that surrealism's interest in nature lay in "the natural production of signs, of one thing in nature contorting itself into a representation of another."[40] But Jules Verne had also commented on the paradoxical nature of coralline forms, and those passages in *Twenty Thousand Leagues under the Sea* were in turn sourced from the work of the French historian Jules Michelet (1798–1874), who is mentioned in Verne's story. In 1861, Michelet devoted a chapter of his book *The Sea* to corals, and titled it "Blood-flower," in which he wrote, "But these animals adorn themselves with a singular splendor of botanic beauty, the splendid liveries of an eccentric and most luxuriant Flora. Far as the eye can reach, you see what, judging from the forms and colors, you take for flowers, and shrubs, and plants. But those plants have their movements, those shrubs are irritable, those flowers shrink and shudder with an incipient sensitiveness which promises, perception and *will*."[41] With these words Michelet evoked the consciousness of coral animals.

In corals, the surrealists—including Breton, Brassaï, and Max Ernst— recognized a metaphor for the complexities of life. Coral reefs suggested the possibilities of realms without borders or boundaries, distinctions or demarcations. They suggested the repressed in modernity: not orderly and rational things but chaotic and eccentric, unpredictable, uncanny, and unconscious things, like the vertigo signified by a Gaudí building in which the architecture was inspired by natural corals. Coral reefs have an uncanny ability to make the animate look inanimate and the inanimate look animate. They have an explosive appearance, as if produced by some spontaneous action. And for these reasons, they became a strategic symbol of nature's inherent surreality, a signifier of freedom, and model for the future of art based in a poetics of defamiliarization. Moreover, Jules Verne had impressed on his readers that corals "embody a sort of natural socialism," and this symbolism, which represents corals as animals who work communally to build the structures of society, explains why, more than any other movement, surrealism, a collective whose members pursued communist revolution, turned its attention to the social model of a living coral reef.[42]

Why did Breton change the geography of the underwater scene? One answer is the relative exoticism of the South Pacific in the 1930s. At that time, the coral reefs of the Bahamas and the Great Barrier Reef of Australia shared the title "Eighth Wonder of the World."[43] For centuries they had been scientific enigmas. But then, in an age of increasing travel and tourism, they became two of the world's most fantasized locations. However, in many instances, the remoter South Seas had more allure. The islands there, being further from Europe, promised something raw and more savage. In the area of popular entertainment and cinema, the relative inaccessibility of the South Seas increased the region's mystique. Consequently, more than once, visual representations of Bahamian islands and seas were passed off to the public as the South Pacific. For example, the American filmmaker D. W. Griffith filmed two silent movies about the South Seas—*The Love Flower* (1920) and *The Idol Dancer* (1920)—in the Bahamas.[44] It was convenient to film the "South Seas" in the Bahamas because the Bahamas was much closer to America yet had all the desired features, including coral reefs, sandy cays, tropical fish, and palms. But psychologically, the South Seas was more attractive to audiences who were seeking in cinematic adventures, the fantasy of escape, freedom, and new beginnings in regions far, far away.

André Breton did not visit Australia but was informed about it through Tristan Tzara's knowledge of Central Australia and Aboriginal song cycles.[45] The Australian region, the cultures of Aboriginal and Torres Strait Islander peoples, and Papua New Guinea were significant to surrealist politics. By changing the geographical location of the coral reef, Breton was able to project on the image his idealized vision of the Pacific as a place filled with imaginary nature and a place of escape from Europe and civilization.[46] The physical geography of the scene he treated as a tabula rasa for reinscription. Once relabeled "the Australian Great Barrier," the image and caption colluded in a form of mutual mimicry—who in 1937 would know the difference between Bahamian corals and Australian corals, and who could tell one underwater location from another?

Once reimagined by a fresh title and different geographical location, the image in Breton's book gave readers a vision of the "edge of the world."[47] The underwater and the Southern Hemisphere were both known as inverted spaces—the first an inversion of land, the second an inversion of Europe. In Breton's idealized vision of the Pacific—which was shared by the surrealists as a group—the Pacific region, especially around Papua New Guinea, possessed a remoteness that was sublime.[48] When the group published "The Surrealist Map of the World," they put the Pacific Ocean at the center of the globe and made Papua New Guinea huge. The sizes of Europe and the United States, on the other hand, were shrunk. Alberto Toscano and Jeff Kinkle explain the motive as primitivist in impulse. It was, they argue, also anticolonialist in its attempt at "belittling imperial Europe and the capitalist USA," and in symbolically challenging white domination of the Indigenous worlds.[49] So it is possible that when Breton renamed the photograph of the Bahamas the "Australian Great Barrier," he intended it an act of *détournement*. Knowing the anticolonialist politics of the surrealists, it is conceivable that Breton intended to hijack attention from the Bahamas as a protest against its colonial system and slave history, the very social context in which J. E. Williamson's life and work had taken shape.[50]

Human figures are absent in the underwater photograph that Breton published in *Mad Love*, but the timing and circumstances surrounding the taking of the photograph, the dynamic context from which this image was extracted, is significant to the wider social and political context of which the photograph was but a part. The photograph shows a mature coral formation that was located during the Field Museum–Williamson Undersea Expedition. Because Williamson produced *Under the Sea* to document the expedition, the entire duration of the event was captured on film. The film shows how minutes after the photograph in *Mad Love* was taken, the same coral formation was chained up by a diver in Williamson's team of African-Bahamian workers. It was extracted from the ocean floor and shipped to the Field Museum in Chicago to become the central feature of the Bahama Islands diorama. Figure 2.5 shows the diver acting up for the camera, just before extracting the coral growth. Behind him are the distinctive shapes and forms of the same palmate coral that appears in Breton's photograph. The same fish, a Nassau grouper, is seen protecting its coral territory. In the image that Breton obtained from the *New York Times*, the diver, who was present for much of the filming, is absent because it was generally the case that when Williamson's photographs of the underwater were published, black workers, who were often behind the scenes smoothing the way for a successful photo shoot, were outside the frame to allow the scene in the final image to appear like a total wilderness untouched by human life. Williamson carefully controlled when the divers were in and out of the picture. When they were represented in the picture, it was usually because the photograph or film demanded a spectacle of savagery or demonstration of Williamson's own sense of cultural supremacy.

Williamson's underwater photograph, reproduced in *Mad Love*, was so affective that it was among the most widely distributed underwater photographs between 1929 and 1937. It was transmitted to a worldwide audience not only by the modern telegraph network but also by traditional print technology. Williamson capitalized on the globalizing media of the early twentieth century, which is why the same image appeared in multiple sources: the *New York Times* (1929); the popular magazine *Our Wonderful*

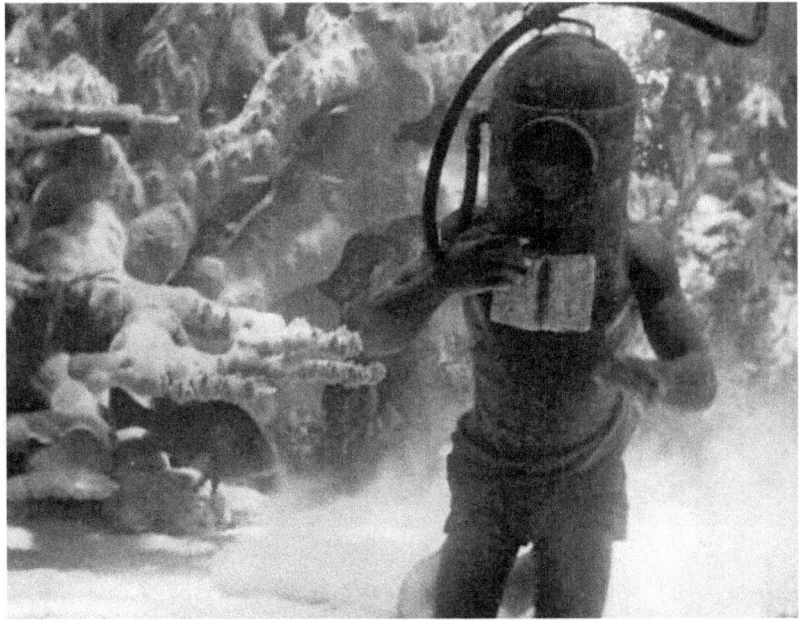

Fig. 2.5. Bahamian diver. Film still. From J. E. Williamson, *Under the Sea*, 1929.
Courtesy of the Field Museum Archives.

World: A Pictorial Account of the Marvels of Nature and the Triumphs of Man
(1931); J. E. Williamson's autobiography *Twenty Years Under the Sea* (1936;
fig. 2.6); and André Breton's *Mad Love* (1937). In *Our Wonderful World*, a
popular set of volumes comprising lively essays and striking photographs
of the natural and man-made worlds, the photograph appeared in an article titled "The Wonder of the Coral Island."[51] Able to cross borders between popular culture, popular science, and avant-garde culture, Williamson's photograph acquired four different captions, two of which placed it
in the two different geographical locations of the Bahamas and the Great
Barrier Reef of Australia. But the caption that offers the best insight into
Williamson's dark imagination is one that accompanied the version of the
photograph reproduced in his autobiography, seen in figure 2.6: "'Where
the huge, loathsome octopus might lie in wait'; A Lovely Setting for Horrible Tragedy." Rather than trying to capture an image of beauty, Williamson wanted to underscore the suspenseful atmosphere of the underwater.

As 1929 drew to a close, visitors to the Field Museum were able to
view for the first time the impressive coral structure (visible in Breton's

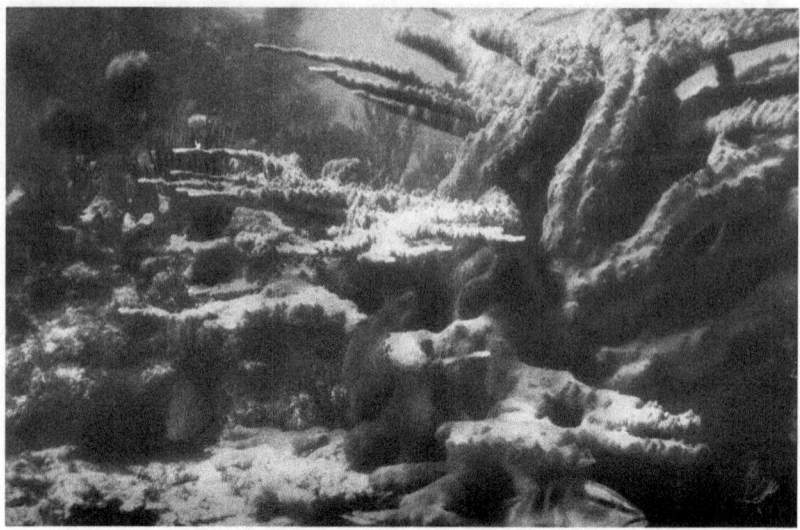

Fig. 2.6. "Where the huge, loathsome octopus might lie in wait." From
J. E. Williamson, *Twenty Years under the Sea*, 1936, 164–65.

photograph) in the museum's "Bahama Islands diorama." In 1953, Breton
visited the Field Museum. He wrote about an affective interaction with a
Sulka mask from New Britain (Papua New Guinea). The mask was alive
with primitive energy, and it epitomized, he said, the "poetry of the sublime."[52] Down a different corridor sat the Bahama Islands diorama showcasing the massive coral structure that had been ripped from the ocean
floor by J. E. Williamson and the Field Museum in 1929, when the underwater coral empire of the Bahamas region was still pretty much intact
but on the cusp of a process of commercialization and exploitation from
which it never recovered. Did Breton recognize the physical coral object
sitting in the Field Museum? If so, did he give it the same regard he gave
the photograph in *Mad Love*? Did the physical object offer the same transformative vision, the "shiver" and poetic dimension that he felt from a
chance meeting with the readymade photograph obtained from the *New
York Times*? Would the three-dimensional coral skeleton have given him
the same sensation of the marvelous?

The circumstance surrounding André Breton and J. E. Williamson's
underwater photograph of the Bahamas draws attention to points that will
be taken up for further discussion in the pages that follow. They include
details about the technology of the photosphere, elaboration on the scene

depicted in the Bahama Islands diorama, and discussion of the methodologies followed during the Field Museum–Williamson Undersea Expedition. The following chapters also return to points raised here about the entanglement of the histories of the Bahamas and the Great Barrier Reef, the hidden significance of black labor to underwater exploration and representation, and an apparent contradiction between the compulsion to domesticate the ocean through metaphors and practices related to aquariums and a desire to symbolize the sea as wild and untamed.

PART II

John Ernest Williamson
and the Bahamas

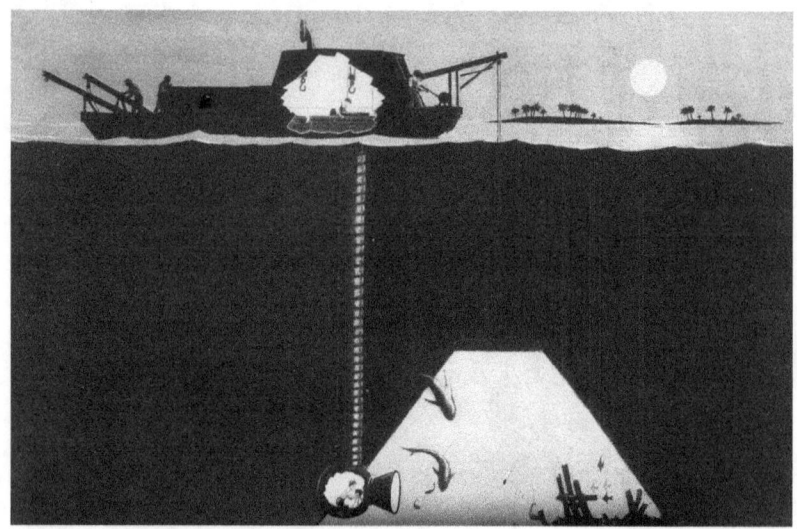

Fig. 3.1. Diagram of the photosphere. From J. E. Williamson,
Twenty Years under the Sea, 1936, 32.

CHAPTER 3

Williamson and the Photosphere

The underwater is a perfect subject for cinema. Few mediums are more dreamlike than the undersea. Few animals are as fluid, and sea creatures that are anatomically very different from animals of the land are surprising and photogenic. Bodies are suspended in fantastic ways free of gravity, and rational coordinates of distance and time are replaced by flickering tones, strange forms, and mysterious spaces. The cinematic qualities of the underwater, and the similarity of the underwater to dreams, explain why a screenwriter in 1956 described *20,000 Leagues under the Sea* (1916), directed by Stuart Paton in conjunction with John and George Williamson, as a movie that "unintentionally borders on surrealism."[1] The screenwriter recognized the same qualities that attracted André Breton to J. E. Williamson's underwater photograph: the uncanny topography of a dream and the suggestion of the unconscious given shape. Appropriated by Breton for surrealist ends in 1937, Williamson's underwater photograph possessed an intrinsic surrealist aesthetic.

In the days of silent movies, and beginning in 1913, John Ernest Williamson produced what he claimed were the world's earliest underwater moving pictures, which he filmed through a glass window in the submersible he called a "photosphere." In astronomy, a photosphere is the part of the sun that radiates light. This optical brilliance is invisible to the human eye, yet it was photographed and fixed as an image on paper in the late nineteenth century by the astronomer Jules Janssen (1824–1907).[2] Williamson named the submersible a "photosphere" because it also enabled the capture of photographic images of sights invisible to human eyes in an underwater world that is normally hidden by the roof of the sea. And when the submersible's artificial lights were turned on, the photosphere also radiated light like the sun (see fig. 3.1).

The photosphere was a masterpiece of modern technology. With a spherical steel chamber four feet in diameter, its interior was large enough for squeezing in two adult bodies. The window, which was made of a sheet of plate glass three inches thick, magnified the underwater scene outside, and through this curved fish-eye-lens window, Williamson's photography enlarged upon nature.[3] The photosphere was connected to the barge *Jules Verne*. It was capable of descending to a depth of 250 feet, although Williamson usually worked in shallower water. Through a flexible tube just wide enough for a body to negotiate with cameras and tripods, he descended to a chamber big enough to sit, stoop, and crouch. He first used the device for filmmaking in 1913 and thought of it as an underwater apparatus that brought together science fiction and military science.

During the First World War, it felt symbolic to Williamson that the photosphere, while based on the *Nautilus* in Jules Verne's story, was not very different from military submarines.[4] He felt part of a technological revolution to utilize the sea and the underwater in ways that could change history, and, as Nicole Starosielski points out, felt as if he too were linked with "naval power."[5] The steel sides of the photosphere were a reassuring barrier to the ocean, because without technological mediation, the underwater was a hostile and dangerous place. The only way to negotiate the medium of the underwater and explore the undersea was by making the body artificial and by using apparatus such as submersibles, diving suits, breathing hoses, goggles, and fins.[6] The underwater was "inner space," as strange and seemingly unknowable as "outer space," and almost impossible to survive in. It was conceived as antihuman and outside civilization.

The world over, people were in awe of the way photographs and films taken from within Williamson's photosphere gave them a perspective on the scale of the ocean yet signaled the ocean's increasing domestication. The submersible was a world in miniature inside the sublime expanse of the sea, yet one journalist imagined the scene from its window as a view into "the most elaborate aquarium you have seen."[7] In other words, the photosphere, like a military submarine, was a sign that humankind was taking charge of the sea. From inside, Williamson cruised the underwater coral reefs of the Bahamas, discovering and revealing its submarine terrains, animals, and secrets. Framed by the round window of the photosphere, the underwater world was revealed to Williamson as a picture, and as truly picturesque. For the most part, nature appeared before him like a painting, or, as he put it, a "field of vision" into which objects, plants, and animals materialized and disappeared.[8]

But the apparatus was difficult to maneuver. The round glass window of the photosphere—the lens of the eye—was at the front of the structure, and the structure itself was not controlled from below. Messages were sent by phone to operators above sea in the *Jules Verne*. And the front-facing position of the window meant that Williamson was unable to see perspectives beside or behind the cylinder. Often the photosphere came to rest in places where plants and corals obstructed the view beyond and so they were "hacked away" to reveal the photogenic vista ahead.[9] The undersea for Williamson was first and foremost a stage for film and photography, and he was driven to orchestrate the best possible viewpoints. In order to frame a more pleasing view through the photosphere's window, he sometimes cleared the seafloor in the same way that terrestrial landscapes are altered for better views. It highlights a conflict typical of the times, a tension that haunts the history of the ocean as a cinematic and photographic site, one that Jonathan Crylen describes as "intimately linked to environmentally unfriendly histories of technology."[10]

For Williamson's underwater photography to be effective, it was essential for the photographed object to occupy the very center of the viewing window, since things at the edges of the window went out of focus. But sea currents and the movement of fishes made central alignment with the camera difficult. The camera was trained straight ahead, and it limited the field of vision. The way James R. Gibson explains the implications of looking at, rather than around, an environment is by comparing the first viewpoint to the "scanning of an object, a page of print, or a picture."[11] It narrows the visual world, gives it boundaries it does not naturally have, and turns the world into information, not experience. It produces a viewer disposition to look and record but not necessarily to develop empathy. Indeed, as Jonathan Crylen points out, in relation to Williamson's cinematography, the frontal viewing position of the undersea, observed through the photosphere's window, corresponds directly with aquarium viewing.

By standing in "dry space insulated from the wet one," the spectator's world inside the photosphere is clearly demarcated from the underwater world in a way that corresponds with the experience of standing in front of the glass of a public aquarium.[12] It produces a sense of emotional distance and remoteness that even the transparent medium of glass cannot rectify. In fact, glass has the paradoxical effect of allowing the sight of the viewer to transition smoothly to the object world beyond, but at the same time puts up a barrier. The dynamics of this is explained by Isobel Armstrong, who writes how glass, however invisible, is a "visible invisibility" that makes it "both barrier and medium."[13]

Aquariums are apparatus that exoticize the world they frame. They emerged as popular sources of entertainment in Western countries during colonial times, which is why Bernhard Klein argues for a direct connection between the modes of spectatorship they cultivate and the "imperial arena."[14] He observes how the "aestheticization of fish and other marine organisms" has a parallel in the objectification and subjugation of the colonized, racial Other. His point resonates when we compare scenes of Western orientalism with the underwater scenes framed by the round window of the photosphere (fig. 3.2). Stills from Williamson's documentary *Under the Sea*, in which marine life is framed by the photosphere's window, compare interestingly with scenes of Eastern life contained within round frames in paintings by the French orientalist and neoclassicist artist Jean-Auguste-Dominique Ingres. In *The Turkish Bath* (1859–1863), for instance, Ingres "captured" scenes of exotic Eastern women that viewers observe like voyeurs through circular frames that imitate the round aperture of a keyhole. The roundel of Ingres's paintings suited the gaze of colonizers, and so did the round window of the photosphere. Framed by the circle of the photosphere's window, the underwater of the British Bahamas was visualized as an oriental and ornamental spectacle of carpets, stripes, tassels, fronds, fans, and feathers. Not coincidentally, when Carl L. Gregory, the cameraman for Universal Pictures' production of 20,000 *Leagues under the Sea*, first encountered the underwater from behind the glass window of Williamson's photosphere, he described it as an "arabesque fantasy."[15] But as with orientalism in Western art, depictions of the exotic Other are more powerful when there is danger. In Williamson's films, the dangerous element is always the figure of a shark gliding into the field of vision.

Gazing from a dry, safe, air-filled chamber through a window barrier

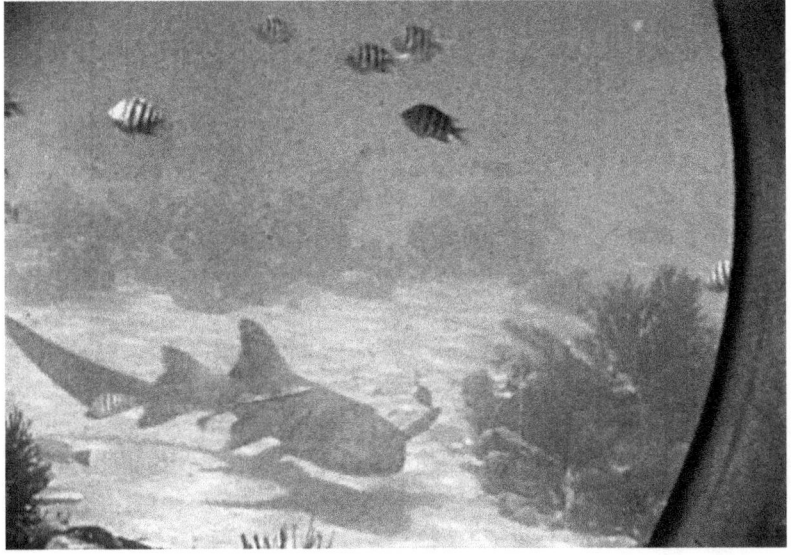

Fig. 3.2. Shark outside photosphere. Film still. From J. E. Williamson,
Under the Sea, 1929. Courtesy of the Field Museum Archives.

at the marine world laid out before him, Williamson always maintained a dualistic attitude to underwater space. Not only did the steel walls of the submersible separate human from nonhuman, they seemed to celebrate the triumph of man and the colonization of the ocean. Each of Williamson's photographs and films were a further transformation of the sea into an object. The photosphere's window resembled the window of the *Nautilus* in *Twenty Thousand Leagues under the Sea*, which Arthur B. Evans describes as a "portal" through which "the heroes—'from the comfort of their own home,' as it were— . . . take in the movie-like spectacle of the outside world."[16] The window, Evans argues, keeps the travelers safe from the outside danger of the Other. But, being transparent, the window also enables the world outside to be observed and studied, and, in the process, tames the wild and makes the outside a possession. This is just the dichotomy that leads Natascha Adamowsky to describe the bifurcation of space created in the submarine environment by the photosphere as "a spaceship-like view of alien worlds and, at the same time, a domestic idyll."[17]

That Williamson modeled himself on Jules Verne's character Captain Nemo is made very plain in *Twenty Years under the Sea*, in which his text addresses an imaginary passenger traveling in the photosphere and tells them, "Please be seated and rest comfortably. You may smoke if you wish. There! The comforts of home! Now, to start on your journey. I draw the curtains aside so that you may see with your own eyes the mysterious floor of the ocean."[18]

J. E. Williamson was an experienced diver, and the frontispiece of *Twenty Years Under the Sea* is a photograph showing him in a diving suit with helmet removed, an expression of wonder and triumph on his face (fig. 3.3). The image reminds us that while in art and poetry the ocean is feminine, in exploration, diving, and photography the ocean is a story about European men and masculinity. But the undersea was a test for even the bravest. Williamson's safe encasement in the photosphere warded off a sense, at times overwhelming, of being a stranger in the inhospitable environment he called "the haunts of creatures of the sea."[19] Williamson argued that the photosphere allowed marine animals to go about their lives undisturbed, inferring that it was them he was protecting. But in truth Williamson had a morbid imagination, and the steel surroundings of the photosphere allowed him to hide from monsters real and imaginary, and enabled him to practice photography and filmmaking as if on land.

J. E. Williamson's father's family was from Scotland, and his mother's, from England. He was born in the English city of Liverpool. He immigrated to Virginia with his family before settling independently in the British colony of Nassau on New Providence Island in the Bahamas. He began his professional life as a reporter, cartoonist, and photographer with the *Virginian Pilot* newspaper in Norfolk, Virginia. In 1913, he and his brother George descended to the bottom of the sea in the submersible their father had built, and they captured photographic images on plates. J. E. Williamson gave credit to his father for producing what he claimed were the first successful underwater photographs, but as the years went by, his father's part in the story disappeared.[20]

In 1913, after J. E. and George had transformed their father's submersible into a photosphere, they formed the Submarine Film Corporation and went into business.[21] Having secured underwater photographs of fish and seaweed at the bottom of Chesapeake Bay, Williamson put them on exhibit at the First International Motion Picture Exposition in Manhattan. He hoped to find a backer for a future motion picture. He printed the images large and hand-colored them for effect, and in his own words his prints were "the greatest attraction of the show," especially with movie stars, financiers, and producers.[22] In 1914, George and J. E. filmed in the

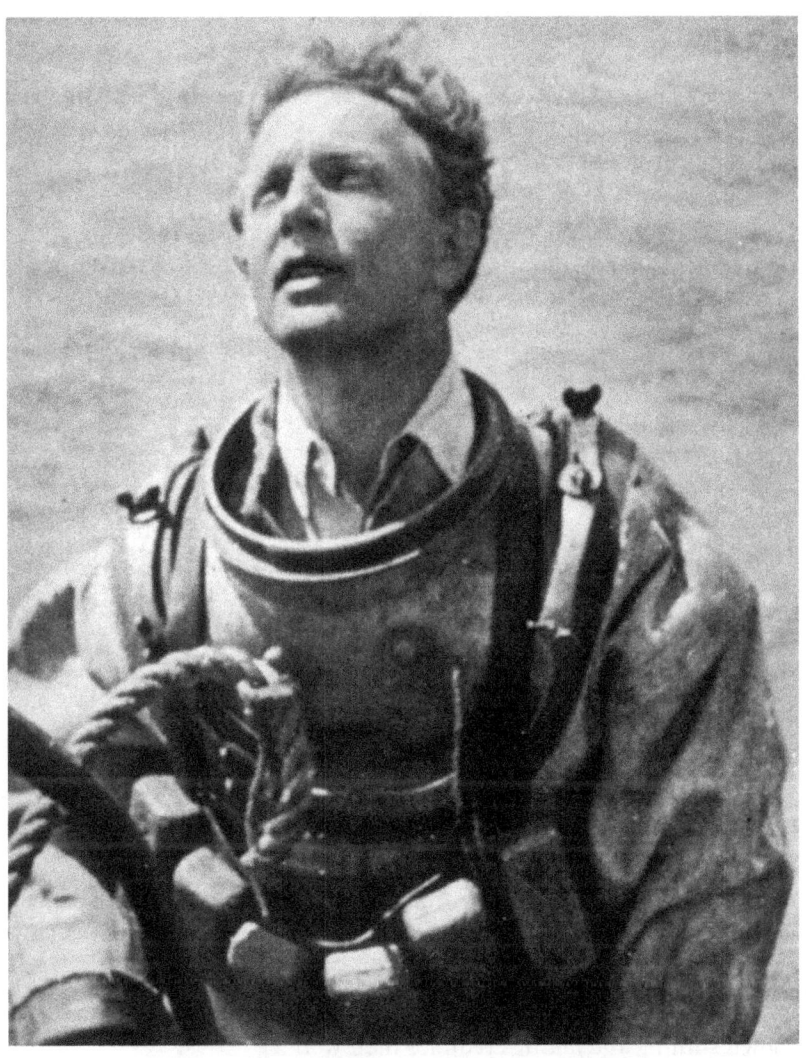

Fig. 3.3. Portrait of J. E Williamson. From J. E. Williamson,
Twenty Years under the Sea, 1936, frontispiece.

Bahamas what they claimed was the first motion picture taken under the sea, and published stills from the event in *Popular Mechanics* (fig. 3.4). The magazine predicted that the undersea "may furnish the background for submarine 'thrillers,' besides which the popular cowboy dramas will seem tame and uninteresting."[23] From there, the idea of the underwater as a stage for human actors helped produce the undersea as a modern spectacle. But in the colonial context of the British Bahamas where Williamson filmed, the underwater also became a space to make a spectacle of the racial Other.

In a British colony defined by barriers between white and black, it was symbolic that the first film Williamson produced in the Bahamas showed three black youths struggling to retrieve pennies from the sea floor. "What a great 'shot' their struggles would make as they snatched the elusive silver," wrote Williamson in his autobiography.[24] Under the pretext of capturing the human body in motion underwater, Williamson also created a portrait of social struggle. In fact, the fetishization and objectification of black bodies in motion would become a hallmark of J. E. Williamson's filmmaking and an expression of the power relations inherent in a segregated Bahamian society.[25] Not only did Williamson racialize the underwater by inferring it as a natural medium for African-Bahamians, but, as Nicole Starosielski shows in her analysis of Williamson's underwater cinema, he "made a spectacle of the native body and fed a widespread cultural appetite for images of the racialized Other."[26]

Eventually J. E. and George fell out and parted company, but not before producing their first feature film titled *The Williamson Submarine Expedition* (1914)—also released as *Terrors of the Deep*, and *Thirty Leagues under the Sea*—which toured first to the Smithsonian Institution in Washington, DC, and then to New York and London. But their most famous collaboration took place when they both helped direct the photography for *20,000 Leagues under the Sea*. The part played by J. E. and George was so important to the film's success that at the start of the film both men are shown smiling and taking credit for their work.

Hurricanes and fires destroyed much of Williamson's film stock, and today his work survives only in fragments, leaving an archive of mostly stills and sections of films that in recent years have been pieced together.[27] Surviving footage shows that his films were a blend of theatre, adventure, and education. He continued to make underwater films in the 1930s, and in 1932 created perhaps the most successful of his underwater docu-

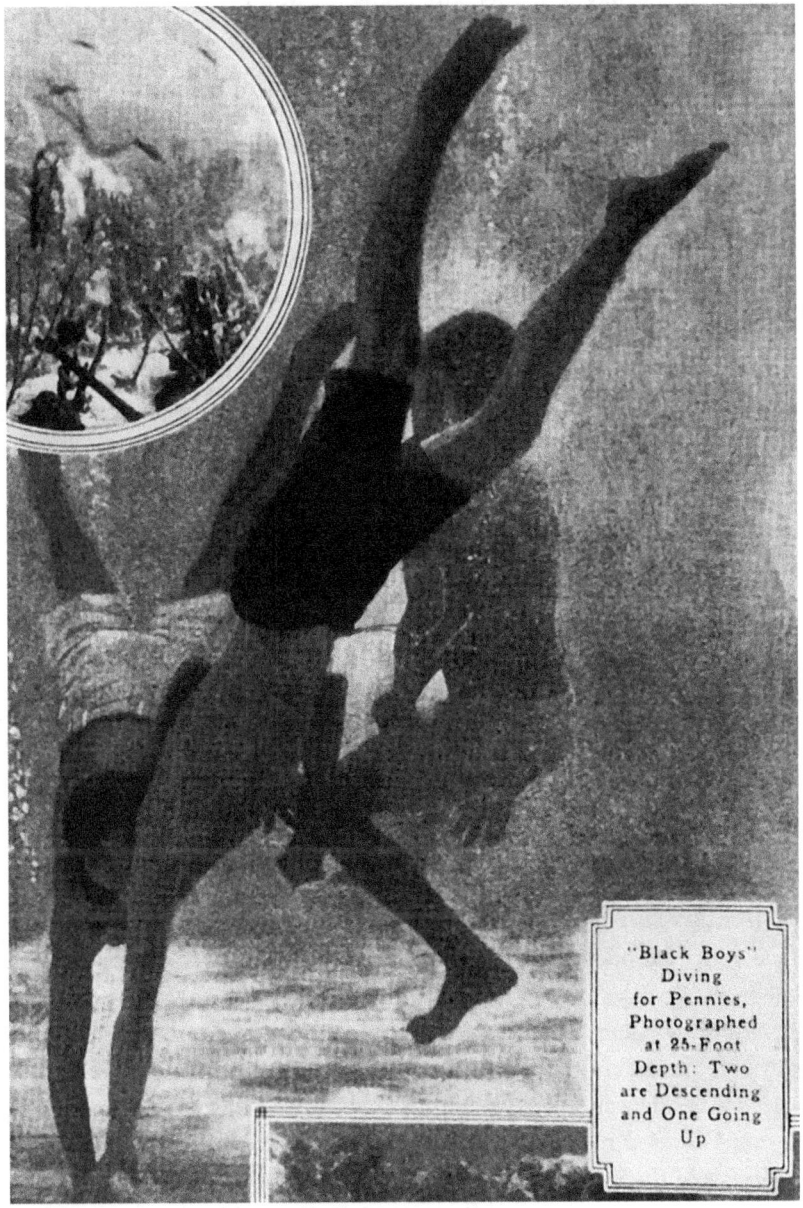

"Black Boys"
Diving
for Pennies,
Photographed
at 25-Foot
Depth: Two
are Descending
and One Going
Up

Fig. 3.4. Diving for coins (detail). From "Taking Movies at the Bottom of the Sea," *Popular Mechanics*, July 1914, 6.

mentaries, *With Williamson beneath the Sea*. It was a film without a narrative, but the way it showed the undersea world unfolding in front of the photosphere's window brought suspense and tension to the experience.[28] Publicity promised audiences an awesome spectacle of "a lost world fathoms below recovered in savage splendor."[29] And audiences were not disappointed. The reception at the Smithsonian was overwhelming, and the press "hailed the showing as 'films that pierced the sea, each picture an absolute revelation.'"[30] On Broadway in New York, one critic praised the documentary as "something never before viewed by mankind."[31] Two years later, the magazine *Popular Science Monthly* proclaimed Williamson the "pioneer photographer of life beneath the waves."[32]

It was important to Williamson's sense of self that he was remembered as an originator of undersea adventures and explorations, as the "first" underwater cinematographer, and as a pioneer of underwater photography. For this reason, he titled the first chapter of his autobiography "The First Pictures out of the Depths." But in the period when Williamson claimed he was the first underwater cinematographer, there was little or no research into the history of underwater photography or diving. When books were finally published in the 1950s, James Dugan described Williamson as "the man who put over submarine photography with the public."[33] In other words, Williamson was first at shaping the popular imagination of the sea through newspapers, news agency photographs, mass-produced magazines, cinema screenings, lecture tours, and journalism. The way in which Williamson differed from others was in the marriage of show business and natural history. His images and films made the underwater a sensation.

To call Williamson's work "submarine photography" was technically true, since he photographed in a submarine, and did so undersea. But the term is misleading because it suggests that the camera was operated in situ underwater. In the early twentieth century, underwater photography and filmmaking was approached in a variety of ways: through a glass window inside a submersible within an air-filled space; with a watertight camera lowered by poles under the water but operated from above water to avoid the photographer stirring up silt; with a camera encased in a watertight container immersed in water along with the photographer; and through the glass fronts of aquariums. Audiences often thought they were witnessing "authentic," revolutionary new underwater photography when in fact the still and moving images they saw were aquarium studies. This was as true for Hollywood cinema as it was for scientific and travelogue films.

Alejandro Martínez explains how the French zoologist Louis Boutan (1859–1934) took the first black-and-white underwater photographs in France in 1893 with his body and camera immersed in water at a time when many of his colleagues, including Louis Fabre-Domergue, used aquariums.[34] Martínez also notes how an Englishman, William Thompson (1822–1879), took underwater photographs in 1856 using a pole.[35] Charles Eldredge relates how an American, Simon Lake, who was contemporaneous with Boutan in the late nineteenth century, achieved results from a submarine.[36] Robert F. Marx claims that Wilhelm Bauer made the first attempt at underwater photography in 1855 and cites Eadweard Muybridge as taking "good underwater photographs" in San Francisco Bay in 1875.[37]

From 1893 on, Boutan's innovations and techniques made news all over the world. People were interested in the waterproof casings he used, the enormous size of the encased camera, his use of heavy glass plate negatives, the fact that he was a diver and wore a deep-sea suit, and the magnesium powder he ignited underwater to illuminate the darkness. That he had taken a sharp photograph at 165 feet was astounding. Boutan was the one who not only told the world how in the depths everything increases in size but also how color there is predominantly green. His knowledge was visceral; his camera sat in the very watery medium he photographed, not separate from it. He was the one who made scientists look forward to learning more about undersea life.

Boutan, in particular, searched for a completely different reality, and he found it when he dived "naked with the pearl fishers of the Torres Straits" with no oxygen supply except his lungs. After that experience, he determined to become an underwater photographer.[38] After diving to the ocean floor and wanting to relay what he saw to others, Boutan was "filled with the desire, therefore, to bring back from these submarine explorations a more tangible souvenir."[39] His early photographs of the underwater were shown in 1900 at the Palace of Optics at the Paris Universal Exhibition, and in the same year he published a book, *La photographie sous-marine et les progrès de la photographie*.[40] The photographs in figures 3.5 and 3.6 show the radical nature of his perspective. He was not content to photograph the sea from above. The perspective he wanted was the embodied, immersed perspective from below, looking through underwater space horizontally.

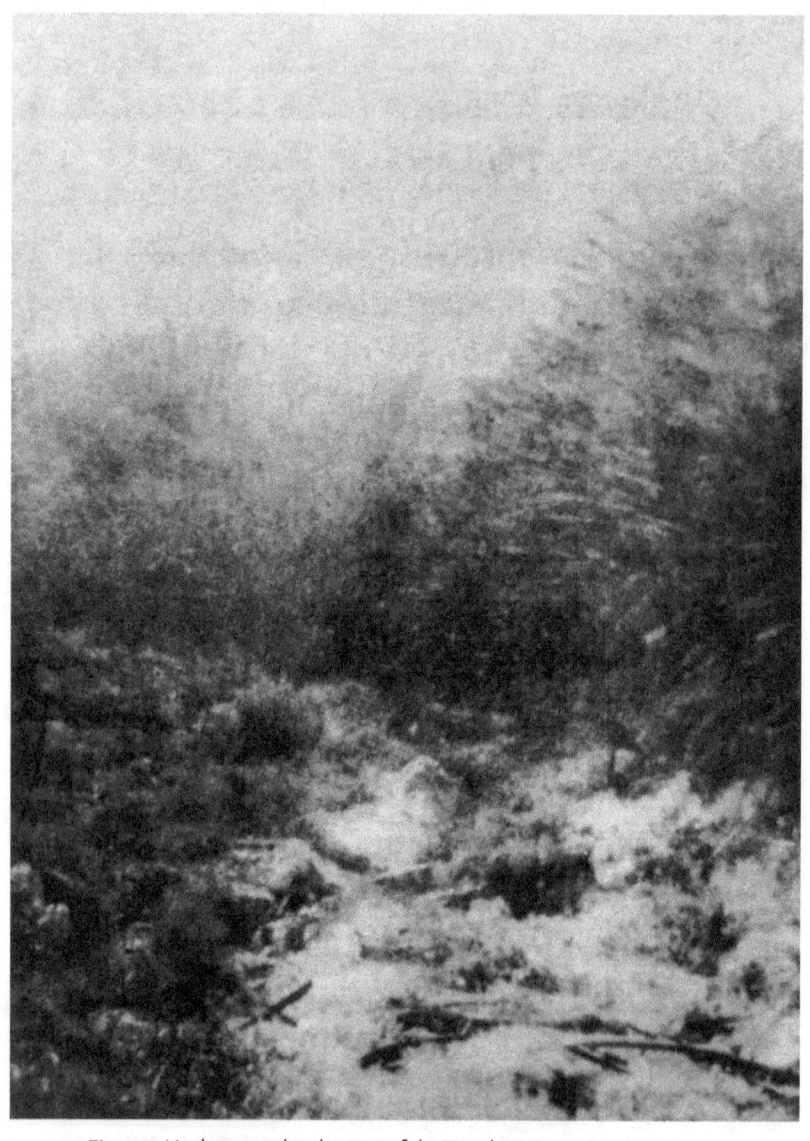

Fig. 3.5. Underwater landscape of du Troc bay. From Louis Boutan,
Underwater Photography, 1900, 188.

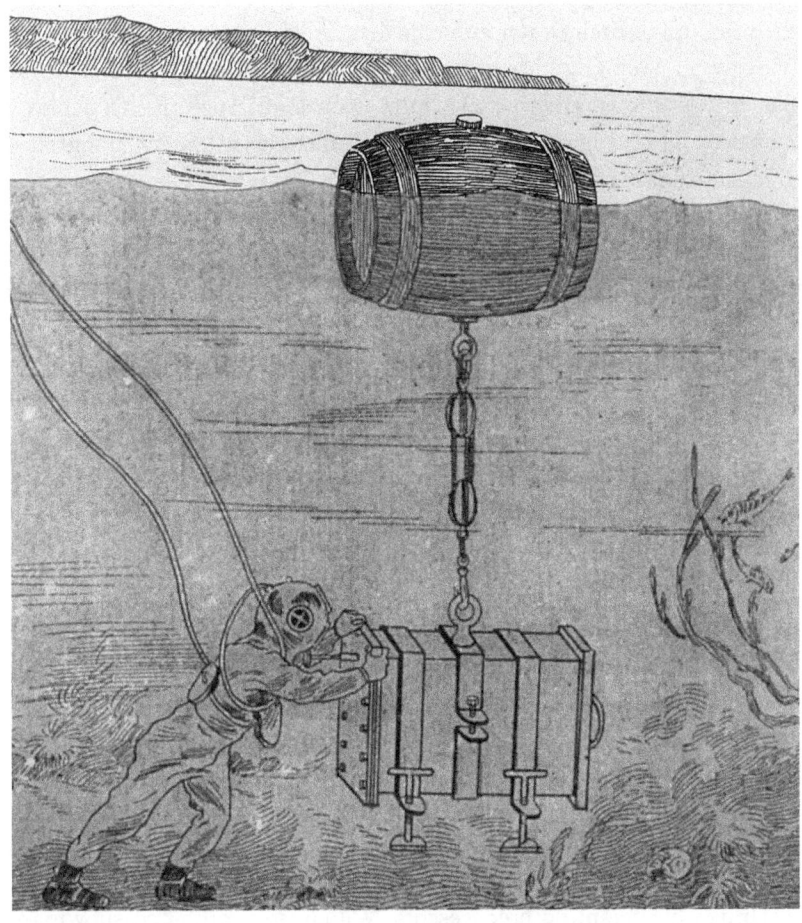

Fig. 3.6. "Manoeuvring the apparatus helped by the float." From Louis Boutan, *Underwater Photography*, 1900, 198.

He wanted to understand the authentic animal in ecological context. But the difficulties of long exposure times of thirty minutes, the need to have a large crew of helpers to maneuver what were very large cameras, and the pressure of underwater currents dissuaded Boutan from continuing underwater photographic experiments after 1900. Instead it was his hope that in the future others would build upon and transform his achievements by designing cameras that would make underwater photography practical and successful.[41]

Many people, including J. E. Williamson himself, claimed that Williamson was the originator of submarine photography, unaware of, or overlooking, the others who had gone before. Did Williamson know about Boutan? If he did, he stayed quiet, even when a journalist, writing about his achievements, claimed that "no one else has ever gone beneath the sea with a camera and brought back a successful photograph of life in the depths."[42]

As the case of Boutan shows, Williamson made inflated claims about being first to photograph undersea, but the photosphere, that feat of engineering that empowered him to transcend the body's limited capacity for the underwater, did give him a unique place in history, particularly in the field of moving pictures. Many people before him had taken films of animals through the glass windows of aquariums. In fact, Gregg Mitman explains that one defining reason for the development of motion pictures was "not for entertainment purposes, but for the analysis of animal motion."[43] Williamson's effort to film animals in motion with a camera operated in the photosphere, under the sea, was also based on the concept of aquarium photography. Nevertheless, his films appear to be the earliest motion pictures taken on the sea floor.

Unlike his scientific predecessors, Williamson combined showmanship, travel, adventure, hunting (with camera, knife and gun), natural history, and storytelling. Through work in underwater film and photography, he hoped to inspire interest and excitement in natural history, to influence the popularization of science, and, equally important, to entertain. To secure entertaining footage of animals in motion, he felt no contradiction in the seemingly opposed approaches and worldviews of the hunter and the naturalist.[44] Entertainment also brought his imagination alive to the myths and legends of the sea. The more monstrous or mythical the animal, the greater the potential to make an exciting film.

When considering Williamson's approach to marine life, Nicole Starosielski observes how "the oceans could simultaneously be revealed to science and still remain in the realm of the supernatural and the exotic."[45] The point is also made by Krista Thompson in writing about Williamson's influence on the history of the Bahamas, and how Williamson, living in that vibrant maritime environment, enjoyed inventing ideas, scenes, and monsters of the sea.[46] It was the explorer and adventurer in Williamson that accounts for why he believed in mythical creatures and legends of the deep. Monster animals of mythic proportions are a feature of his writing. When asked about the unexplained disappearances of divers from sea vessels, he responded, "Who can say that these mysteries of the sea, these unexplained tragedies, were not brought about by gigantic devil fish, veritable Krakens?"[47] Mythical monsters were creatures that had never been caught, but to Williamson this simply meant they were invisible, not that they didn't exist.

When Williamson began a career in underwater photography, the ocean had only just come to light as the site of evolution for life on earth: it was a prime symbol of savage nature, and because of this, embodied immense psychological intensity. The concept of devolution, of gigantic animals that never evolved and that remained hidden in the planet's most remote regions, was a major point of interest in popular culture. The theory went that sea creatures were more likely to evolve in shallow water, where competition for survival is keenest, whereas the creatures of the deep water remained as they always had: fearsome, grotesque, big, and prehistoric.[48] When the Williamson brothers were employed in 1915 to direct underwater scenes during the making of 20,000 *Leagues under the Sea*, J. E. Williamson was especially excited to be involved in a scene in which a diver is caught in a giant octopus's winding tentacles. He chose the location carefully: "A rocky reef overgrown with marine life, with gigantic coral-trees whose roots formed dim mysterious grottoes and caverns, an eerie spot where the sinister shadows of great sharks moved menacingly through the undersea jungle; where huge morays twisted like giant serpents among the sea fans, where crabs lifted great claws and where, in any of the murky caverns, the huge loathsome octopus might lie in wait for its prey."[49]

It was a mutation of Charles Darwin's theory of evolution that explains why the craze for monsters and monster hunting gripped the world in the early twentieth century and why popular culture, particularly Hollywood, sought out stories of nature gone awry. When *King Kong*, directed by Merian C. Cooper and Ernest B. Schoedsack, was released in cinemas in 1933, it unleashed a new excitement about animality, devolution, and the concept of nature as a place of struggle, change, and unpredictability. The very idea that man is descended from apes was proof that evolution occurred through freak occurrences, but the possibility that colossal animals still roamed the remoter regions of the planet took this concept to another level. The Loch Ness Monster, for example, generated considerable excitement in the 1930s. What freaks of nature, what leviathans were lurking unseen in the undiscovered depths?

Given Williamson's fascination with underwater "monsters," it is perhaps unsurprising that in 1934, after the *Scotsman* newspaper reported a strange sight on Loch Ness, Williamson joined other adventurers and went looking for the Loch Ness Monster.[50] Marina Warner points out in relation to anxiety, the sea, and sea monsters: "Knowing your monsters can help contain them."[51] Williamson took the photosphere with him in the hope of getting a close-up photograph. He was sure that the Loch Ness Monster was a baby squid that had entered the lake through a channel connected to the sea, then grown to gigantic proportions. This he explained in *Popular Science Monthly*, where he also said he had determined to "settle the controversy as to whether the creature is a sea serpent of fabulous size, as some observers insist [or] a large whale, as naturalists have tentatively identified it from long-distance photographs."[52]

Not much in the way of additional detail was released about the expedition, including whether Williamson made a successful descent of Loch Ness in the photosphere. But in 1937, he told the *Los Angeles Times* that while he failed to see the Loch Ness Monster, he had spoken to enough eye witnesses to be convinced of its existence, and he was standing by his argument that the creature was a huge squid "with arms more than sixty feet long." In his view, "It is just as unreasonable to doubt the existence of hitherto unknown animals in the sea as to conclude that the earth is the only planet among probably billions that is inhabited. A few miles down in the sea there is far greater darkness than on the blackest night. No scientist knows what sort of huge creatures may inhabit this blackness."[53]

Part of Williamson's logic was similar to thinking about nature and the limitations of human perception and knowledge put forward by surrealists. Just because something can't be seen by a human being doesn't mean it isn't there. An animal, like a leaf insect, can be so well camouflaged that it easily goes undetected by the human eye. Therefore, argued the surrealist John B. L. Goodwin, who knows what monsters live invisibly in the phenomenal world.[54] These ideas, which Williamson also subscribed to, were current in the 1930s and 1940s, and were in part fed by the increasing interest of movie makers in science fiction and cryptozoology. But in truth, the fear of monstrous animals of the sea haunted Williamson during his ventures underwater. Not even the clarity and vibrancy of Bahamian waters could ward off the imagined gaze of the animal Other: a monstrous shark, a giant octopus, or any hitherto unimagined freak of nature.

Williamson sought out the charismatic Bahamas because Bahamian waters were famous for being the clearest in the world, the islands were sun-filled, and under the surface of the sea was a photogenic, alternative reality of coral reefs. In 1889, biologist Francis Hobart Herrick (1858–1940), writing in the *American Naturalist,* described how at Nassau "the crystal water is like a lens, and the sandy bottom like a white screen, which reflects and diffuses a soft light through the ocean depths."[55] The Bahamas was also legendary for palmate corals (*Acropora palmate*) that grew thick branches twenty feet high in towering fields described as "forests of living stone."[56] The very fact of the reef's presence, despite its predominant invisibility, stimulated a desire to reach it and an impulse to photograph it underwater. From the mid-nineteenth century to the mid-twentieth century, there was a shift in consciousness in relation to coral reefs, a change from wanting to experience one firsthand by standing on an island or rowing in a boat to wanting to get under the surface of the water to see the reef from below and eye to eye. Gazing down at a coral reef through crystal clear water involved a specific vertiginous experience. But getting beneath the surface and seeing the ocean from within promised a different kind of disorientation, a new perspective on nature. In 1927, the zoologist William Beebe, who conducted much of his marine research in a submersible in Haiti, close to the Bahamas, described how he looked up from underneath the water among coral reefs and saw the surface of the sea from a completely different viewpoint. It looked to him like a "green, wrinkled, translucent ceiling cloth."[57]

"In the tropics, of course," wrote J. E. Williamson in 1933, "it is easier to obtain good photographs, for the water is not only transparent, but iridescent."[58] The water changes color as the angle of viewing changes, and as the angle of light changes. If it had not been for the vividly transparent waters of the Bahamas, and the towering fields of living corals, Williamson's photograph that André Breton appropriated for *Mad Love,* and that

Breton found symbolic of the process of life's formation and destruction, might never have been made, nor would the Stuart Paton movie *20,000 Leagues under the Sea*. It was the quality of light in the Bahamas that determined Williamson's career as a photographer. As much light as possible was needed to capture clear impressions of the underwater on film. Williamson's awakening to the glassy environment of the Bahamas was stimulated by the commercial potential of what seemed to be a perfect union between photography and tropical light in bringing images to the screen. A calm tropical sea was like a lens focusing the eye on the depths below or, if under the water, on the transparent expanse above. But when the sea was rough, it turned from liquid sapphire to "black oil."[59] In bright sunlight Williamson could photograph without lights at a depth of sixty feet, but when the sky and sea changed, he always ascended to the surface.[60] There was no point staying below to photograph the ocean in disturbed conditions—the whole idea was to make objects as visible as possible while making the milieu surrounding them invisible.

In the early twentieth century, there were some parallels between the light and clarity of the Bahamas and the modern city, particularly in the way the clear waters of coral reefs offered a nature-based metaphor for modern life: spectacular, visible, and true. Attraction to transparency at the Bahamas was similar to the appeal of the transparency of the glass façades and windows in modern cities. As Kevin Robins explains about the poetics and politics of modern vision: "Nothing would remain invisible, nothing would remain outside the field of vision."[61] The analogy of crystal clear Bahamian water resonated with modernity's obsession with transparency, a desire that Anthony Vidler describes as one that seeks out "spatial penetration and ubiquitous flow of air, light and physical movement."[62] But while light and transparency were defining features of the physical Bahamian landscape, the country's social landscape was shrouded in a dark history and troubled economy. Lucayans were the original inhabitants of the Bahama Islands; they were expert divers and underwater spear fishermen until they were enslaved by Spanish colonizers to dive in Haiti and Cuba in the sixteenth century, and eventually these coastal people were wiped out.[63] Europeans who colonized the Bahamas established a plantation economy, but it collapsed in the nineteenth century, after which the industry of the sea based on pearls, shells, and sponges, and that depended on the diving skills of African-Bahamians, also began to decline due to depletion through overfishing.

When J. E. Williamson settled in Nassau, the tourism industry and American investment were on the rise. His own underwater exploration and filmmaking in the Bahamas was part of the shift toward a tourist economy, and in a broader sense part of a new phase in the commercialization of Bahamian seas. After so much publicity surrounding the photosphere and Williamson's ingenuity, the American film industry was attracted by the sensuousness of the islands and their proximity to the United States. Meanwhile, in the field of science, museums sensed a unique opportunity to further the natural history of the sea by deploying Williamson and the photosphere, something Williamson encouraged because it brought the nature of his work closer to the seriousness of science.

Despite Williamson's taste for fiction and adventure, the possibility of gaining respect within scientific circles was also appealing. It resulted in his involvement with two museum expeditions. First, he teamed up with the American Museum of Natural History in New York to gather material for an underwater diorama, and later he was employed by the Field Museum in Chicago to build the Bahama Islands diorama, an exhibit that everyone hoped would be the greatest underwater coral reef diorama ever built.

CHAPTER 4

The Field Museum–Williamson Undersea Expedition

The commanding coral edifice that filled the frame of the underwater photograph that J. E. Williamson published in the *New York Times* in 1929, and that André Breton reproduced again in *Mad Love* in 1937, was collected during the Field Museum–Williamson Undersea Expedition to Sandy Cay in the Bahamas in 1929. After the coral was pulled from the seafloor, and following a process of bleaching and cleaning to remove living polyps, the skeleton was placed in the Field Museum's Bahama Islands diorama and put on public show in late 1929 (see plate 2). Williamson would later remember the search in the photosphere for a perfect coral specimen and the sight that met his eyes: "A giant palmate, a golden tree beneath whose spreading branches we came to a stop. Rising majestically from the bulbous trunk of the tree like hundreds of outstretched arms with upturned hands, the majority of the branches reached valiantly toward the life-giving ocean current."[1]

After removing the coral branches from the water, Williamson was later bewitched to see them "cleansed and bleached to snowy whiteness," ready for exhibition as edifices.[2] Corals, for Williamson, were beautiful in death as well as in life. As dead things, they gained a new sense of beauty through pure form. This way of thinking suited perfectly an era obsessed with hunting, collecting, skinning, preserving, and mounting animals for display, and a culture in which exotic animals were routinely killed but also given "an afterlife" as exhibits.[3]

Without the technology of the photosphere to enable Williamson and the Field Museum to descend into the undersea and observe nature first-hand, the diorama would never have been built. Being a "habitat diorama," the methodology behind its construction was to go into the wild to photograph, draw, trap, kill, skin, and preserve animals. It was an approach better suited to research for terrestrial habitats, and in the 1920s posed an almost insurmountable problem for the undersea study of marine animals. With a submersible, though, the underwater could become accessible. Williamson knew that in the area of marine natural history exploration, the photosphere gave him an edge. In later years, he would describe the sense of responsibility he felt about the "tremendous assignment of bringing the bottom of the sea to Chicago."[4] But in 1929, he felt nothing but optimism for the new era of underwater museum exhibit that the photosphere would help usher in. In August 1929, he wrote to the Field Museum's president, Stanley Field, to impress upon him the uniqueness of the Bahamas expedition, which, he said, would take them into "the almost unexplored depths," and he urged Field to publicize the photosphere as "the instrument that has opened the way to present and future research and discovery under the sea."[5]

The deployment of Williamson's photosphere ensured that the Field Museum–Williamson Undersea Expedition was not the usual oceanographic adventure of the type in which scientists such as Alfred Russel Wallace, aboard the British Navy ship HMS *Challenger* in 1873, stood on deck to net specimens and dredge the seafloor. This one would take place underwater. It would be recorded in still and motion pictures, and these would be used as reference material when constructing the diorama. They would also serve as educational aids for future lectures on the science of the undersea. Williamson and the Field Museum's director, Stephen C. Simms, agreed that an article in the *New York Times* would leave the public in no doubt about the innovative nature of their venture and its value to science. Publicity photographs were arranged for release through the *New York Times* and its photo agency, Wide World Photos, the same agency in whose stock photographs André Breton would later discover an image taken during the expedition that suited the theory and aesthetics of surrealism.[6] But before this, in correspondence with Simms in February 1929, Williamson was resolute that any pictorial record of the expedition must carry credit lines to ensure that, in the future, documentation was duly attributed to the Field Museum–Williamson Undersea venture.[7] Wide World Photos honored that demand when it stamped Williamson's name, and details of the expedition, on the back of the image that Breton acquired. But, as we have seen, the credit line was crossed out before the image was published in *Mad Love* as an anonymous photograph of the Australian Great Barrier Reef.

The museum hoped to build one of the greatest dioramas the world had seen, one depicting an underwater scene in a coral cavern with a group of sharks as the main motif. Although Williamson's role was to guide the expedition and take charge of the undersea collection of specimens, he had, from the beginning, an opportunity to shape and influence the design of the diorama display. If they were to work effectively and efficiently on location undersea, they needed a clear sense of the final design. Williamson assessed the museum space and told Simms that he was enthusiastic about "a major group of unique composition from the Sandy Cay reefs and life there."[8]

The scientific objective with the Bahama Islands diorama, which was destined for the new Hall of the Ocean Floor in the Field Museum, was to utilize taxidermied animals and background scenic painting to replicate the biogeography of the region and show authentic species relations. A "habitat diorama" is one in which taxidermied animals are shown in settings that simulate their true ecological context.[9] From the outset, though, well before the expedition began, the Field Museum planned the diorama as a macabre dramatization of the savagery of nature.

Rather than conceiving of it as a coral reef diorama with the inclusion of sharks, the Bahama Islands diorama was conceived as "a great shark habitat group" with corals. The Field Museum's taxidermist and zoological preparator of fishes, Leon L. Pray (1882–1975), described the concept: "A major group containing sharks as follows: Hammerhead, Tiger, 'Man-eater' and some smaller kinds. These fish to be mounted in a group circling a coral pocket from which the blood of an injured fish or eel is diffusing thinly in a cloud up through the water. With characteristic sea-floor accessories."[10]

A ready-made nature consisting of generic ocean objects, such as sand and shells, with a narrative based on life-and-death struggles among the animals of coral reefs was integral to the original concept. It was also a concept that preceded the expedition's empirical research, meaning that the nature it intended to display was already predetermined.

The diorama was planned as an animal hunting scene to give audiences in the Midwest who had never seen the ocean, or visited the tropics, the sensation of being under the sea and of witnessing the natural order of life in which sharks, the sea's top predator, give chase to prey. How to show the gruesome truths of the food chain, and survival of the fittest, but also faithfully reproduce animals in their natural setting? Those were challenges that faced the development and construction of the Bahama Islands diorama in 1929. But the public in the 1920s had an insatiable appetite for shark imagery, and the fact that tropical Bahamian waters were known as "shark-infested" further fed a deep fascination with sharks, death, deviancy, and cruelty.

In fact, the diorama's final representation of sharks, depicted as they circle and roll, bears a noteworthy resemblance to Winslow Homer's painting *The Gulf Stream* (see plate 3). In the late nineteenth century, Homer was a regular visitor to Nassau, and one of his greatest fascinations was with something that Patti Hannaway termed the pervasive "shark-versus-man theme."[11] Winslow Homer's paintings *The Gulf Stream* (1899) and *Shark Fishing* (1885) are among the most famous Western interpretations of this theme. In *The Gulf Stream*, a frenzy of sharks circles a lone black sailor adrift on a boat. He is embroiled in a battle with nature, one that Homer said he had witnessed in the region, namely "the continual struggle of men against the elements, including the battle for survival between man and shark."[12]

After 1906, when Homer's painting was purchased by the Metropolitan Museum of Art, it became an American national treasure. Its connection with the Field Museum diorama can therefore be seen as both direct—in the sense of inspiring the Field Museum to create its own powerful visual imagery of sharks—and indirect due to the pervasive fascination with shark imagery in American culture. Although no human figure features in the Field Museum's Bahama Islands diorama, the exhibit is nevertheless a variation on the shark-versus-man theme. The glass barrier of the diorama demarcates human from animal, and celebrates the triumph of man over animal by exhibiting shark bodies that had been subject to capture, dissection, and scrutiny, and then rebirth as sculptures.

As powerful visual experiences, both Homer's painting and the Field Museum diorama had great potential to influence and shape a terror of sharks and fear of the sea. Stephen Quinn notes that the experience of viewing habitat dioramas can be so persuasive that understandings and preconceptions of "nature" are easily influenced by illusionistic exhibits that appear like real moments frozen in time.[13] The challenge for audiences visiting a museum of science is to observe the scene objectively and dispassionately. But in the case of the Bahama Islands diorama, this was an impossible ask: fiction and science meet in an explosive way, and the diorama exploits to the fullest a pervasive social and psychological fear about sharks on the attack.

The concept behind the Bahama Islands diorama was in large part a response to public and scientific fantasies and fears about the facts of nature. When the Field Museum–Williamson Undersea expedition party left Chicago in March 1929 bound for Nassau, one team member was Leon Pray. Within a month, Williamson had taken Pray into the photosphere undersea among the coral reefs to observe sharks up close. Williamson then wrote to Simms and described how Pray had experienced "the thrill of seeing a couple of big sharks with their companion fishes, the Jacks, cruise menacingly around us."[14] It was a vision they preserved for the diorama's final design: the sight of three menacing sharks of different species with remora fish attached by suckers to their sides.

The horror of Bahamian sharks made those sleek, fast animals popular game for hunters, and J. E. Williamson, being both hunter and naturalist, killed, as well as filmed, a great many sharks in his lifetime. The varied geography of the Bahama Islands allowed Williamson a range of depths and shallows in which to photograph. The best place to find sharks was in the deeper zones, such as the Tongue of the Ocean between New Providence Island and Andros Island. Williamson's chapter on sharks in *Twenty Years under the Sea* stands as one of the earliest, lengthy, firsthand observations of shark behavior. But he was not a disinterested observer: sharks were vicious, untrustworthy, and dangerous. Mostly he felt horrified by them and called them "wolves of the sea," "man-eaters," "ravenous beasts," "the enemy," "frenzied demons," and "cold-blooded cannibals." When the view from the photosphere was pierced by the black dorsal fin of a shark, he photographed the sight as if witnessing the scene of a crime. A parallel on land was the underworld of a city. The undersea also harbored shady life and suspect characters, of which the shark and the octopus were Williamson's most feared.

Despite this fear, a defining moment in Williamson's career was when he swam underwater during the filming of *The Williamson Submarine Expedition* (1914) and killed a shark with a knife in front of the camera. Killing sharks was blood sport, and the joy of killing a display of masculinity and bravery despite the reality that Williamson preferred the protected space of the photosphere to immersion in the sea. The scene in which he had the camera turned on himself as he stabbed a shark—a shark that had been attracted to his boat by a dead horse and animal blood poured in the water by Williamson's crew—takes place very quickly.[15] But by completing the act, Williamson had delivered on a promise to a financial backer that he would secure on film a fight between a man and a "man-eater."[16]

J. E. Williamson claimed that his intentions during the Field Museum–Williamson Undersea Expedition of 1929 were purely "scientific."[17] But clearly a quest to find the truth sat in tension with a desire to excite the imagination. He was delighted at the thought that the public would visit the Field Museum and "look with wonder" at the undersea diorama and what it revealed about nature.[18] Yet he exhibited deep ambivalence toward ocean wildlife and marine habitats. Stalking and facing sharks was a desire not just to kill but also to draw a line once and for all between the human and the animal in the context of what Carrie Rohman refers to as Darwin's "catastrophic blow to human privilege vis-à-vis the species question."[19] Rohman's investigation of the question of the animal in modernity exposes the anxieties that underpinned the uncertain status of the human in light of Darwin's theory that man is also an animal. Mirroring these anxieties were Leon Pray's incorporation of a "man-eater" shark into the Field Museum's vision for the Bahama Islands diorama, and the influence of Williamson's evident unease with the animals of the sea in the diorama's final design. The plans for the exhibit were as much about human-animal relations and the social construction of nature as they were about natural history, and they put the possibility of objective science in doubt. The point is further underscored by Donna Haraway's observation that habitat dioramas are visions in which animals are "actors in a morality play on the stage of nature."[20] The scene that Pray and Williamson devised for the Bahama Islands diorama is a moral drama between a supposedly merciless and cruel predator on one hand, and a weak and defenseless prey on the other. How much was the public able to learn about nature and ecology, and the ocean's ecosystem, about the natural behaviors of animals, and the growth of corals, when the exhibit came into existence to sublimate desires and obsessions?

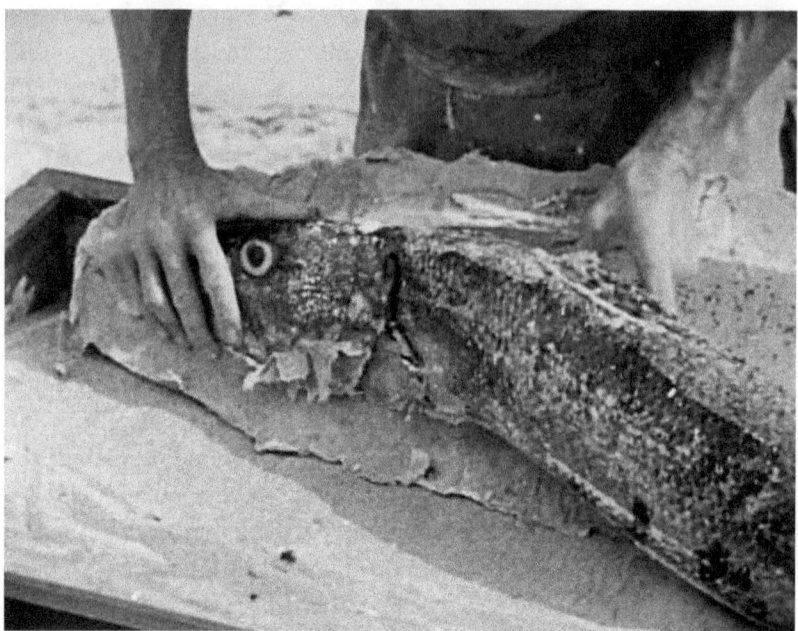

Fig. 4.1. Leon Pray molding a barracuda. Film still. From J. E. Williamson,
Under the Sea, 1929. Courtesy of the Field Museum Archives.

On Sandy Cay, Pray's job was to receive freshly caught fish, such as barracuda and sharks, mold them on site, and ship the molds to Chicago for
taxidermy, at which point the skins were stretched over mannequins and
painted realistically (fig. 4.1). For the final shark mannequins to appear as
lifelike as possible, the animals had to be fresh when molded, so they were
brought to Pray alive, strapped to the side of a boat, then dragged onto a
beach before being finned and tailed and finally set as molds.[21] At Sandy
Cay, Pray molded hundreds of sharks and other fish, and Williamson described him as "a scientific artist" who brought imagination to the ob-
jectivity of natural history.[22] To Williamson, Pray was a gothic figure, not
unlike an undertaker. It was striking to Williamson that Pray always ate
the meat of the fish he defleshed, which he prepared in a ritualistic way,
refusing to wash the flesh or touch it with his own hands.[23] It especially
intrigued Williamson to see Pray "open up his coffin-like boxes, laying
out a profusion of cutlery—long sharp scalpels, knives of all shapes and
sizes, pincers and scissors of various designs, an endless assortment of

glittering surgical instruments. Resting on the operating-table were tropical beauties of fish-life waiting to be offered up as a sacrifice to science."[24]

In the end, though, once the animals had been turned into mannequins for the diorama, none of Pray's labor was apparent. Pray had been apprentice to Carl Akeley, who, as a naturalist, filmmaker, explorer, and the taxidermist responsible for the extraordinary dioramas in the Akeley Hall of African Mammals in New York's American Museum of Natural History, was legend. Akeley's taxidermied animals seem alive and only momentarily frozen in time. Before taking up his position in New York, Akeley had been employed by the Field Museum, where he met Pray.[25] Art in the service of science is how habitat dioramas have been described.[26] But they are also described as "the art that conceals art."[27] What they try to make invisible is human presence in their construction; what they try to convey is an unmediated experience of nature.

A distinguishing feature of natural history museums in the 1920s— including the Field Museum in Chicago and the American Museum of Natural History in New York, but also the Queensland Museum in Brisbane and the Australian Museum in Sydney—was the urge to build undersea, tropical coral reef dioramas.[28] The growth of tropical tourism was partly responsible, as was Charles Darwin's treatise *The Structure and Distribution of Coral Reefs* (1842), in which he proposed that reefs are formed by the earth's subsidence. But it was the aesthetics of coral reefs, in particular the magical colors of corals but also the suggestion of danger among reef grottoes, coupled with a desire to colonize the underwater, that put coral reef dioramas at the forefront of museum displays in the 1920s.

The Field Museum Expedition was not the first time the photosphere had been put to use in the Bahamas for museum science. It was the second. From 1924 to 1926, Williamson was employed by the American Museum of Natural History to collect corals from the Andros Reef for a new diorama in the Milstein Family Hall of Ocean Life.[29] The man responsible for organizing the expedition was the museum's curator, Roy Waldo Miner (1875–1955). It was Miner who undertook a fact-finding mission that employed J. E. Williamson in the first stage of the process of collecting a significant assemblage of corals for the American Museum.

When completed, the diorama occupied the vertical space of just over one floor of the American Museum. Visitors can still observe the Andros Reef from above as well as from below (fig. 4.2). If standing on the floor

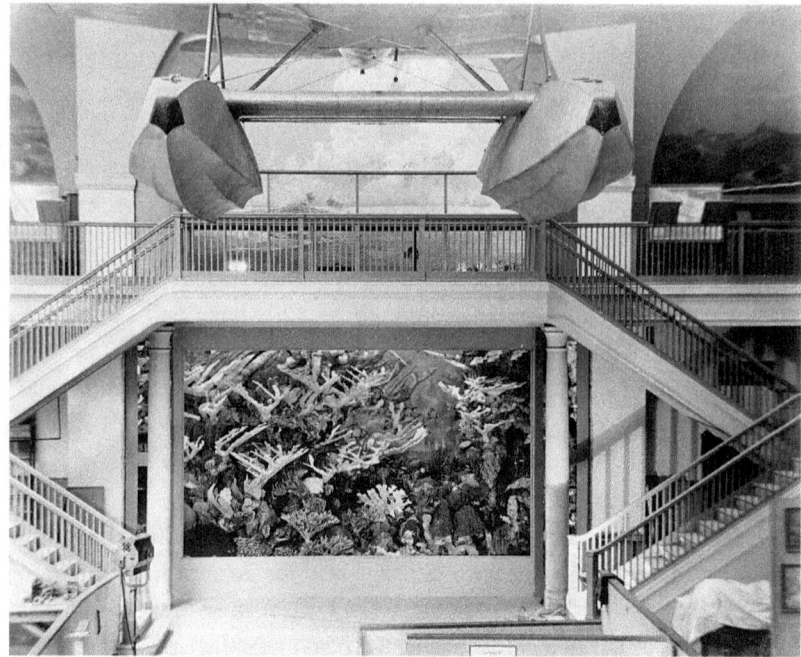

Fig. 4.2. Coral Reef Group, Hall of Ocean Life, 1934. Image 314541,
American Museum of Natural History Library.

above, they see a familiar terrestrial view looking out across a picturesque
sea to islands and shorelines, to rocky outcrops, and flora and fauna. A
protrusion of weathered limestone hints at what lies underneath the blue
surface of the sea—a lurking, invisible coral reef. Tension builds as viewers descend to the floor beneath where the undersea perspective of the
reef, from eighteen feet below the surface, greets them. The encounter is
decentering; it is difficult to comprehend that such a dark and primitive
nature was concealed by the tranquil and picturesque view above. To see
the diorama from below is to witness a seething tangle of coral branches,
rocky caverns and grottos, hidden spaces, strange animals, and dangerous potential. It shows a world that humans do not belong to, and seem
unwelcome in. It does not try to endear to audiences the underwater, or
a submerged coral reef. Rather, it feeds the morbid imagination. It bears
out the words of the British sailor and adventurer Frank T. Bullen when,
in writing about the sea in 1909, he warned that the ocean "reveals to the
voyager no inkling of what is going on below its mobile mask."[30]

As if to strengthen and further extend the unsettling, eerie encounter with the underwater seen in the Andros Reef exhibit, and the disorienting sensation of observing it as if from the seafloor, the American Museum built a second underwater diorama next to the Andros Reef. It opened at the museum in 1941, and belongs in style and concept to both the Andros Reef and the Bahama Islands diorama in Chicago. It shows another equally intense, primitivizing vision of the undersea. Titled "Pearl Divers in Coral Reef, Tongareva, French Polynesia" (see plate 4), the diorama reveals, again, a dark cavernous space; this time the scene is peopled, but in a way that adds to the fear.

In the Polynesian diorama, two Pacific Islander free divers go in search of pearls. Miner had descended the reef with Indigenous divers and was impressed by the way they held their breath for minutes at a time.[31] By observing them he learned about diving. He saw divers wearing tortoise-shell goggles, and from them learned the significance of putting a layer of air between the eyes and the sea. But when his observations were transformed into diorama mannequins, the goggles reflected a demonic light, and the divers' strangely elongated fingers and limbs and their fixated expressions became a source of alienation for viewers. There is an inference in the Tongareva diorama that these are native peoples who long ago adapted to the underwater environment by becoming amphibious.

Because the Polynesian diorama also looks like an aquarium, the human figures seem merely one odd specimen among an array of strangely shaped fish and weirdly formed corals. They dive toward and work on the seafloor in a place that symbolizes the past, the site of evolution and distant time. Viewers are left with the sense that the underwater is terrifying, and that the people who are at home in it are more animal than human.[32] In future pages, we will see how this racialization of the underwater also came to figure in writing and films by J. E. Williamson, as well as in the reviews his films received.

In the early twentieth century, Miner stressed the importance to science of the horizontal and underwater view of a coral reef. It was this view, he argued, that enabled the modern scientist to obtain new knowledge and a true understanding of ocean nature. The same position was held by Heber A. Longman, director of the Queensland Museum in Brisbane. When in 1928 the Queensland Museum built a diorama of the Great Barrier Reef, Longman said he was impressed by the innovations of the Andros Reef diorama, which made news in Australia for the way it enabled visitors to view it "from above, below, or through the sides."[33] He also argued that to see a coral reef from below and to look at it horizontally was the modern way to observe such a phenomenon of nature. Looking from above was outdated; looking from below the surface, horizontally, was scientific. What was needed, he said, "was some sort of submarine which would descend to a depth of about 20ft. and stay there awhile. If that could be achieved one could get a much more intimate knowledge of the life history of these beautiful marine growths than was at present possible."[34] Longman had no such device, but in 1924, Miner, with J. E. Williamson's help, was able to attain a horizontal view and take photographs by descending in the photosphere. By 1934, though, the photosphere was redundant for scientists—including Miner—who by then had learned deep sea diving and acquired new waterproof cameras to aid their science.[35]

J. E. Williamson therefore served museum communities in the United States at a critical juncture for marine science: the period after scientists, photographers, and explorers had determined to pursue the horizontal viewpoint of the sea underwater, but when it was still necessary to do so from a submersible, especially since the undersea was still a place of fear; and the period before a shift in thinking to the pedagogical, experiential, and aesthetic benefits of immersing the human body itself in the underwater with the aid of innovative camera technologies that were waterproof enough, and film innovations that were fast enough, to photograph in dim light.

Williamson also served museums at a critical time before a decline in the popularity of diorama technologies, which were replaced by cinema. As John Miller explains in relation to Carl Akeley's work in the Field Museum and the American Museum, the advent of moving pictures meant that dioramas were "poised in the 1920s on the brink of a terminal decline."[36] Being protocinematic, the Andros Reef and the Bahama Islands dioramas invite the audience to experience the duration of time, and this they do by forcing audiences to walk. At the American Museum, walking the diorama involves descending a flight of stairs, while at the Field Museum it involves a horizontal passage from one end of a large glass case to the other, witnessing in the process an ever-changing spectacle of schools of fish and coral forms. As the viewer walks the Bahama Islands diorama, events are enacted by each group of animals, from tiny fish to major predator; their actions unfold like a film. Everything in this large-scale work is laid out for the spectator like a screen.

To that end, the glass front of the display had two purposes: to imitate the transparency of water and allow the eye to glide through space as if the viewer and viewed share the same environment, yet at the same time to act as a barrier, separating human and animal, and protecting viewers from the cruel and primitive scene before them. The Bahama Islands diorama wants viewers to feel they are on the ocean floor, but also to feel relief that they are not part of that world. Instead there is a heightened awareness that they are safe witnesses to a vision of predatory animals in a state of nature fighting for survival. Horror at the brutality of nature, alleviated by pleasure in occupying a safe distance, is the aesthetic response orchestrated by its design. This appeal to nature as a source of anxiety and pleasure has a parallel in the late twentieth century in Damien Hirst's diorama of a four-meter tiger shark suspended in formaldehyde. By gazing into the open jaws of the animal displayed in *The Physical Impossibility of Death in the Mind of Someone Living* (1991), the viewer finds an unnerving mixture of horror and delight, although here it differs from the Bahama Islands diorama because it is not accompanied by the pretext of elucidating scientific knowledge. However, because Hirst challenges audiences to face their nightmares, the work compares quite well with the Chicago diorama. The makers of both exhibits sought the largest possible specimens, and for what purpose if not the psychological impact of the animal?

Today, the Chicago diorama stands as a relic of a bygone era, when the urge to attain mastery over nature led to what Richard Conniff calls the "mad pursuit" of animals for collecting, classifying, and stuffing. Paradoxically, it was driven by the melancholic belief that due to colonization, urbanization, and the march of modern progress, wildlife was rapidly vanishing.[37] The American bison, New Guinea birds of paradise, and the New Zealand huia bird were three victims of overhunting for sport, food, trophies, jewelry, fashion, and taxidermy. In the name of science, they were hunted down by the very museums that sought to preserve them, but it was not seen as exploitative. Today, though, with the fundamental shift toward an ecological perspective, to sustainability and environmental protection, the Field Museum acknowledges the once competing demands of "admiring, studying and preserving" the natural world.[38]

The Field Museum–Williamson Undersea Expedition, and the earlier expedition staged by the American Museum, resulted in colossal numbers of animals, including fish and corals, being extracted from the Bahamas for displays. More than three tons of coral were taken from the ocean during the Field Museum–Williamson Undersea Expedition. By the end of the expedition, fifteen enormous crates of *Acropora cervicornis* (staghorn coral) and *Acropora palmata* (elkhorn coral), corallines and fans, many types of sharks, large and small fish casts, shark jaws, and small pickled fish, had been shipped. The scale of extraction for the New York exhibit was so vast that it gave some people reason to marvel at the ingenuity of Man.

For instance, when the *Illustrated London News* reported on Roy Waldo Miner's expedition to the Bahamas in 1925, it published a photograph of bleached palmate coral skeletons in store for transportation to New York, and boasted that it was "probably the largest specimen of coral ever taken from the sea-bed."[39] In 1936, Frances Jenkins Olcott published an excerpt from Miner claiming that "forty tons of coral [were] ripped from the heart of a hundred-mile submarine forest" for the diorama in New York.[40] Eighty years later, Trevor Norton would note about the American expedition: "There would soon be more of the Andros reef in the United States than in the Bahamas."[41] The impact on the environment was substantial. After the 1920s, there was a devastating decline in coral cover in the wider Caribbean area.[42] The environmental situation prompted Stephen C. Quinn, of the American Museum of Natural History, to acknowledge that the Andros reef in the Bahamas has been so ecologically degraded that the coral reef diorama in New York "could never be duplicated now."[43]

The message about the brutality of nature, a main theme of the Bahama Islands diorama, is a magnificent diversion from the brutality that brought the exhibit into existence. Camouflaged behind the spectacle of sharks and fish and animals supposedly going about their natural lives, but frozen in time as if by magic, is the grim reality of an exhibition of dead animals that met a violent end. Later Williamson would recall how difficult it had been to secure the sharks.[44] As it turns out the whole gruesome event was documented in *Under the Sea* (1929), a four-reel film now in the collection of the Field Museum.[45] The film is predominantly a combination of two genres: nature films and hunting films. From the photosphere window the viewer watches fish and animals in their natural habitat, but the viewpoint moves from below sea to above when sharks are captured, killed, and molded.

How were the sharks caught? By a team of African-Bahamian workers who were required to bring them in alive. In *Under the sea*, the conflict depicted between man and nature is presented as it is in Winslow Homer's painting, as a specific conflict between shark and black man. The film enacts a principle of animality that Carrie Rohman interrogates in relation to Darwinian evolution. Writing about the crisis of identity that Darwin's theory that man is an animal provoked, Rohman observes how "the displacement of animality onto marginalized others operates as an attempted repression of the animality that stalks Western subjectivity in the modernist age."[46] This principle can be read into Homer's painting, in which man and animal are forced into a dangerous and intimate relationship. Peter H. Woods proposes that in the painting, the man's struggle is more than a struggle with mortality, it is a racialized struggle, and the painting an allegory of slavery and servitude.[47] A similar reading can be made of *Under the Sea*. While Williamson's purpose was to document the process of collecting specimens for the Bahama Islands diorama, and to educate about the science of ecosystems, the film, like the diorama, was produced by social systems underpinned by racism and distinctions of class arising from a history of slavery and colonization in the Bahamas. *Under the Sea* provides insight into the people whose labor, methods, processes, and tasks enabled the Bahama Islands diorama to get built, revealing the behind-the-scenes relationships and social dynamics, the treatment of animals, and, to large extent, a brutality toward African-Bahamians evidenced through the violent gaze of the camera—all of which is absent from the final diorama display.

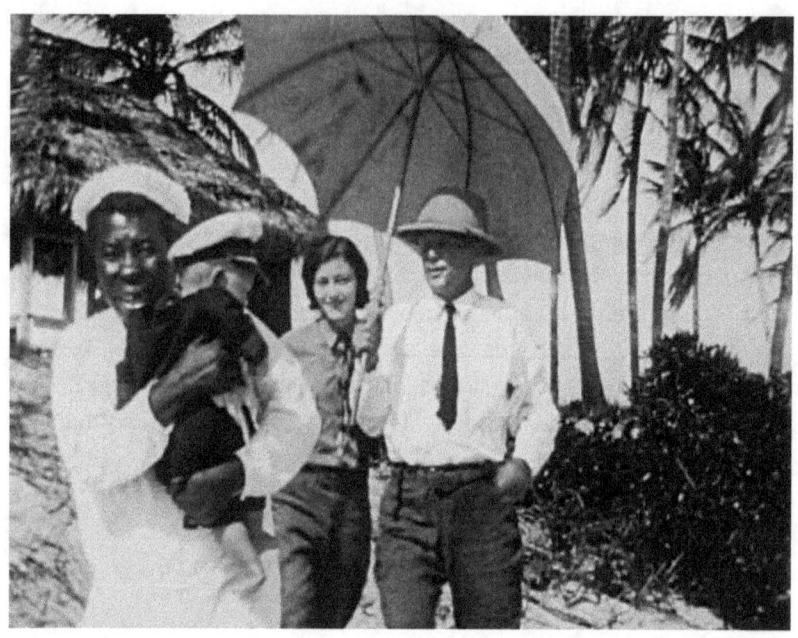

Fig. 5.1. J. E. Williamson at home in the Bahamas. Film still. From J. E. William-son, *Under the Sea*, 1929. Courtesy of the Field Museum Archives.

CHAPTER 5

Under the Sea

The Bahama Islands diorama was planned in February 1929 as a shark habitat group, meaning it could be realized only if sharks living in Bahamian waters were hunted, collected, and prepared as mannequins for the exhibit. *Under the Sea*, a documentary filmed mostly by J. E. Williamson, records the Williamson–Field Museum Undersea Expedition, and devotes considerable footage to the "native crew," Williamson's term for the group of men who were employed for a range of jobs, in particular to dive, and to capture fish and sharks. Williamson claimed the divers were relaxed around sharks. In his mind, there was something natural about "native" man against beast, and it was something he had maintained for years, certainly since publishing an article in the *New York Times* in 1916. There he described how effortlessly black divers were able to kill sharks, suggesting that man and shark were natural combatants and enemies. He explained how the divers had perfected a technique in which they "dive deep and come up under their prey. One quick upward thrust with a sharp knife does the trick. It is only necessary to hold the knife firmly and allow the swift-swimming shark to rip itself open."[1] The way Williamson represented the relationship between "native" and shark in writing and filmmaking relates to a point that Paul Gilroy makes about race-thinking and how blackness is often produced as being "between animal and human."[2]

Williamson's life and work with the underwater for film companies, museums, and tourism would have been unthinkable without black labor. To make these operations succeed, he relied entirely on a labor force of divers, carpenters, and servants. He was one of a minority population of whites who lived in Nassau. In his safari helmet and tropical whites, the standard uniform of white people in the colonial tropics, he presented a cheerful picture of power (fig. 5.1). Michael Craton and Gail Saunders explain that despite emancipation and the growing non-white middle classes of the British Bahamas, there was a strict separation in Bahamian society of whites and nonwhites of African descent.[3] The legacy of this is very clear in Williamson's writing, including the autobiography from 1936, and in his filmmaking.

The workers Williamson employed were descendants of slaves who had arrived in the Bahamas by different routes and circumstances. Some came from Africa to British colonies in the seventeenth and eighteenth centuries, not only to the Bahamas but also the West Indies. Others had escaped to the Bahamas from areas of enslavement in the United States in the 1830s, including from Virginia, where Williamson had once lived. A third group was descended from slaves brought from the United States to the British Bahamas to work on plantations by exiled loyalists after the American Revolution.[4] After abolition, enslavement gave way to servitude, and it was a culture of black Bahamians in servitude that determined the quality of life enjoyed by J. E. Williamson.[5]

In his ability to control machines, especially the futuristic photosphere, and to control Bahamian workers, Williamson was the very image of Euro-American colonialism. The work he assigned to divers was often dangerous. Sharks were unpredictable and potentially unsafe, but so was decompression sickness by nitrous oxide poisoning. Often the divers were included as actors in Williamson's films, and occasionally they ascended from the ocean floor too quickly. In 1916, for example, during the making of 20,000 *Leagues under the Sea*, the divers surfaced after a dive and "suddenly fell upon one another like maniacs, fighting desperately. [They were] struggling, utterly crazed by the exhausted chemical."[6] When Williamson needed shark-fighting scenes, he almost always put black divers in front of the cameras. He exploited the way they would take on the job in exchange for meager sums of money. In 1930, the *New York Times* reported, "Mr Williamson said he found some natives who are willing, for a small remuneration, to risk their lives in encounters with sharks while motion pictures are being made."[7]

Jules Verne, in a passage in *Twenty Thousand Leagues under the Sea* about shark hunting, drew attention to exotic places where "Negroes don't hesitate to attack sharks," but also to the myth that race makes some people immune to attack.[8] Sharks were a dark force in Bahamian cultural memory. For one, they were linked with the history of slaves brought from Africa in the eighteenth century. Two accounts by Marcus Rediker and Alan Rice put this story in perspective. Rediker relates how the threat of sharks was used to terrorize slaves on the Atlantic voyages and deter them from escaping.[9] Rice describes how some slaves were used as examples to others and were dropped into the sea, where they became "food for the sharks."[10]

The writing of the Atlantic emerges in the social intersection of the history of ocean space with slavery, diaspora, and colonization. It is something Helen M. Rozwadowski details in her book on oceans and the fusion of environmental history and the science and technology of the sea.[11] Consequently, any analysis of J. E. Williamson's film *Under the Sea* should be considered in context of the broader inquiry into the Atlantic's history, including contemporary interrogation of the human-animal binary in colonial history, a field in which John Miller specializes.[12] Miller explains how resemblances between native peoples and animals "was a common trope of colonial discourse, forcefully asserting the supposed gulf between colonizer and colonized and opening a range of violent and repressive possibilities for colonial rulers as racial others are emptied of their human status."[13]

Through visual representation in films and photographs, J. E. Williamson sought to naturalize the place of nonwhites in savage, tropical environments and succeeded in turning shark fighting into entertainment and sport.[14] These are reasons why *Under the Sea* is much more than a record or document of a scientific expedition. It is also a visual stage on which race-thinking is performed, and where conceptions of history, nature, the animal, and social power relations are enacted. *Under the Sea* was conceived as a natural history film and as a document of an expedition in order to fulfill the contract that Williamson had with the Field Museum. But a variety of social and cultural narratives and stereotypes are relayed throughout the film, not only about nature and natural history but tensions between white and nonwhite worlds, human and nonhuman animal orders, and colonizers and the colonized as well. Each stereotype conforms to nature as "primitive, bestial, and corporeal," the states and conditions that Kate Soper argues "reflect a history of ideas about membership of the human community and ideals of human nature . . . through which Western humanity has constituted and continuously re-thought its own identity."[15]

In *An Eye for the Tropics*, in which Krista Thompson explores a history of the British colonial islands of the Bahamas and the West Indies, she points to the double bind for African–Bahamians. On one hand, they were valued by whites, who saw them as a picturesque part of nature and as such "completed the scene of tropicality."[16] On the other, they were placed in the category of nature, not culture, which was dehumanizing. White people in the Bahamas drew a line between their own claim to civilization and the place of nonwhites. So, when J. E. Williamson wrote how the African-Bahamian divers he employed were "nearly as amphibious as human beings can be," it was a sign he believed they were inferior to whites for being closer to nature.[17] Similar primitivizing beliefs are found in whites' accounts of Pacific Islanders. In a book on Oceania, for example, Paul D'Arcy notes that in history, "numerous observers were struck by how comfortable Islanders were in the water, and often described them in terms usually reserved for marine creatures," including the collective noun "shoals" used for groups of Indigenous swimmers at the Marquesas in French Polynesia.[18] These examples also offer a possible explanation for why the divers in the American Museum diorama "Pearl Divers in Coral Reef, Tongareva, French Polynesia" appear amphibious (see plate 4).

Williamson referred to the team of men he employed to fight sharks as "submarine gladiators."[19] It was his way of acknowledging manliness. There was no question that he admired the athleticism of the Bahamian crew. Krista Thompson observes how Williamson enjoyed tracking the "performance of black agility" with his camera.[20] He also admired speed and agility in sharks, so in the combined image of the motion of "native divers" and sharks in the underwater Williamson recognized extra excitement, extra savagery. But the term "gladiators" stereotyped the workers as brutes, and it reconnected them to a history of slavery when, in ancient times, black gladiators were also slaves.

Occasionally in *Under the Sea*, the camera faces J. E. Williamson, but primarily the four reels of film record the months of work at Sandy Cay and other islands in the Bahamas during the collecting expedition for the Bahama Islands diorama. The footage includes the camera surveying the floor of the ocean underwater from the vantage point of the photosphere, but some sections of the film concentrate on what amounts to an ethnography of black labor, with Williamson training his camera on workers above and below water as they went about collecting and shipping marine material for the Chicago diorama. In some scenes they descend the underwater as free divers, in others they appear before the photosphere's window as cyborgs in diving suits connected to the *Jules Verne* by air hoses, fighting undersea currents in lead boots.

Williamson's film shows how the men were not only skilled at diving but also at fishing and handling sea craft ranging from two-hulled canoes to modern maritime machinery. But it also reveals the social division between their manual labor and his own supervision. Many scenes strive to characterize Williamson as a benevolent manager, yet the contemporary viewer cannot fail to see in these scenes Williamson's own sense of racial superiority. John Barnhill has pointed out that one feature of colonialism is the way colonialists not only believed in their own superiority but also rationalized their belief by arguing that the colonized received economic and social benefits and it is this that comes through about Williamson in *Under the Sea*.[21] In truth, Williamson did not entirely trust the African-Bahamian men who worked for him. At times he wondered if they would turn against him. For instance, after disappearing into a darkroom to develop cinematic film, he instructed the crew to lock the door, but noted, "A queer idea flashed through my mind. What if they did not respond to my signal for release? This could happen. Hadn't I damned them to Hades a hundred times, driven them like a slave driver, and, in emergencies, treated them like dogs, though I loved them all for the children they were by nature! Supposed, by some queer twist in their make-up, they turned on me now."[22]

If anything, *Under the Sea* illustrates perfectly why it was only a matter of time before the people of the Bahamas who had been portrayed as inferior, characterized as objects, missionized, patronized, and coerced into labor would seek independence and self-determination.

The first reel of *Under the Sea* begins with a shot of the Field Museum in Chicago and then the camera scans the seascape of the Bahamas, where in the distance a white-fringed coral reef with palms is visible. Then a moving pointer on a map of the United States tells the reader that the film will record an expedition from Chicago to the Bahamas, and the viewer is left with a clear idea of the geographical location of the Bahamas, where the rest of the film will be made. The absence of sound puts emphasis on visual signs. A white couple, J. E. Williamson and wife Lilah (1895–1992), drink tea while a black nanny plays with their child on the beach (fig. 5.1). Within a matter of a minute, therefore, the film introduces the audience to the man who will lead the expedition, and to the structure of Bahamian society.

A barge appears with a team of African-Bahamian men. They lower the photosphere into the water, and J. E. climbs down the connecting tube with his baby and wife. The camera lingers on the vessel's machinery, especially a concertina tube that delivers people to the photosphere's chamber under the water. The emphasis of the film is on ingenuity, labor, and machinery. As J. E. disappears with the baby, a diagram comes up to show how the photosphere is designed to sit on the ocean floor with a bright light to illuminate the undersea. Next thing, we gaze at the dreamlike spectacle of schools of small fish flashing past the window of the photosphere like a film screen. The sight is magical, and it makes sense of something Jonathan Burt recently wrote about marine animals on film: they are "objects of pure fascination on screen [and] a spectacle of the marvelous."[23] J. E. was obviously behind the camera, since his wife and baby are silhouetted against the photosphere's window. The film cuts to a diagram of a fish skull, and a man's hands introduce a three-dimensional fish skeleton to demonstrate how a fish's mouth works. At this point the emphasis is on nature study. As the photosphere is moved along the floor, new scenes come into view, showing us the range of corals, grasses, and marine life underwater near New Providence Island, the Bahamas.

The film is slow moving until a shark appears, cruising the ocean bottom with remora trailing behind (see fig. 3.2). The dorsal fin is glimpsed as a menacing shape gliding behind rocks and weeds. The shark is shown to be machine-like: methodically and precisely scouring the floor of the ocean. Here the window comes into its own by separating the homely order of life inside the photosphere from the otherness of the sea outside. It is a particularly modern form of vision in which the absence of participatory involvement leads, as Martin Jay explains, to a widening of "the gap between spectator and spectacle."[24] Next thing a remora is shown above sea, on a boat. It is put on display for the viewer by a man, probably Leon Pray, who rotates the fish in his hands and demonstrates how a large sucker enables the fish to cling to the skin of a shark. Then the camera cuts back to scenes of the ecological diversity of the coral reefs underwater. It is an abrupt, rather than fluid editing style, of the type often found in silent wildlife documentaries, and it creates, as Eric Barnouw points out, an artificial "film time" that also allows for the manipulation of events.[25] A scene of fish clustering around a coral head that has been removed by divers and suspended in watery space is utterly intriguing for its representation not just of the collective behavior of animals but also of the behavior of animals manipulated by human presence. Whether Williamson intended it or not, the scene provokes the viewer to shift attention from the human world above water to the impact of human life underwater and to question whether the behaviors of wild animals on screen can ever be classified "authentic."

The second reel begins with a scene of a free diver, one of the crewmen from the barge, who holds his breath underwater outside the photosphere's window, trying to engage the attention of the baby with a starfish. Next, a diver with helmet and air hose arrives and takes the helmet off, also to amuse the baby. The film again cuts to a scene of a shark, cruising the sea floor, adding more drama and narrative to what is otherwise a slow-paced nature film. The shark is portrayed as a menacing presence; what will happen to the diver? Williamson then cuts to a scene above water in which a man on a boat hacks into the body of a small shark and flings the meat and blood overboard, then back to the underwater to watch a live shark take hold of the carcass of the mutilated shark, shake it vigorously, and eat it. The message is: sharks are cannibals and have no conscience or morality.

A man drags a stingray on board a boat, shows the audience its underside, and teases it, even though the animals is still alive. A split second later we see the ray dead, and a mold being taken by Leon Pray, who then prizes a barracuda out of a mold and shows the viewer how big its teeth are. Down in the photosphere, J. E.'s wife is sketching and painting the scene outside. The fish are teeming. They are all shapes and markings. A huge "tree" of palmate coral comes into view silhouetted against the ocean. It is instantly recognizable as the coral structure in the still photograph that André Breton published in Mad Love, the same distinctive growth of coral that is now in the Bahama Islands diorama in Chicago. This section of the film is shot close to the moment that Williamson took the still photograph that Breton obtained from Wide World Photos. Captured in moving images, the dimension of time adds another layer to the history of both the still photograph and the coral growth. Seen in moving images that bring the past back to life, it seems tragic that the coral colony was not left where it grew. The in situ viewpoint is melancholic; the vision of healthy, growing corals a stark reminder of their imminent death. The camera focuses on a Nassau grouper guarding its den. When a diver wearing lead boots comes too close, the grouper tries to menace him. Is that the same grouper that became a stuffed specimen in the diorama? The diver chains up the coral growth for removal. The seabed is left empty where the coral had once been, and the fish who lived in and near the coral are left behind, swimming in circles.

Collecting for the diorama continues, and in the third reel the audience witnesses the difficulty of working underwater. The currents are strong, and the divers find it difficult to stay upright. We get a sense of the precariousness of the life of the diver as the air hose of one snags on coral. The divers make every effort to show us the enormous scale of the coral growths by standing beside and underneath them. They are dwarfed by the scale and look as if they are in "fairyland." Back in the photosphere, there are shots of J. E. sketching the coral formations that the diver outside is perusing, sketches that will act as scientific aids for the final design of the diorama. There are scenes of a worker on the beach stuffing sponges into sacks as cushioning for the corals that will be shipped to Chicago. J. E. inspects the work and helps out. The men are building a wooden cradle to retrieve the largest coral structure, and the film then lingers on the protracted processes of labor, of pulling the structure from the water using pulleys and chains rigged up on a beach.

We see the shape of the coral structure emerge from the deep and appear just below the surface. One man uses a water glass to have a better look (fig. 5.2). The coral structure is broken, and they throw it away. The next structure they bring up is intact. At this point the camera focuses on the men, as if the film had always been intended as an ethnographic study. It shows Williamson's team of black workers eating, laboring, running, building crates, and measuring. They seem to be working quickly, knowing the camera that is filming them is also surveilling their work. The final scene of this reel is surreal. The giant coral structure is hauled by chains and slowly emerges from the water like a ghost: this is the same coral structure in Breton's book, and the same skeleton exhibited in the Bahama Islands diorama.

For the final reel, J. E. is in front of the camera as the coral is prepared for shipment to the Field Museum. He measures out the length of the largest specimen, and breaks off an oversized branch so it fits in the crate. The scale of the structure is awesome (fig. 5.3). It has been bleached white

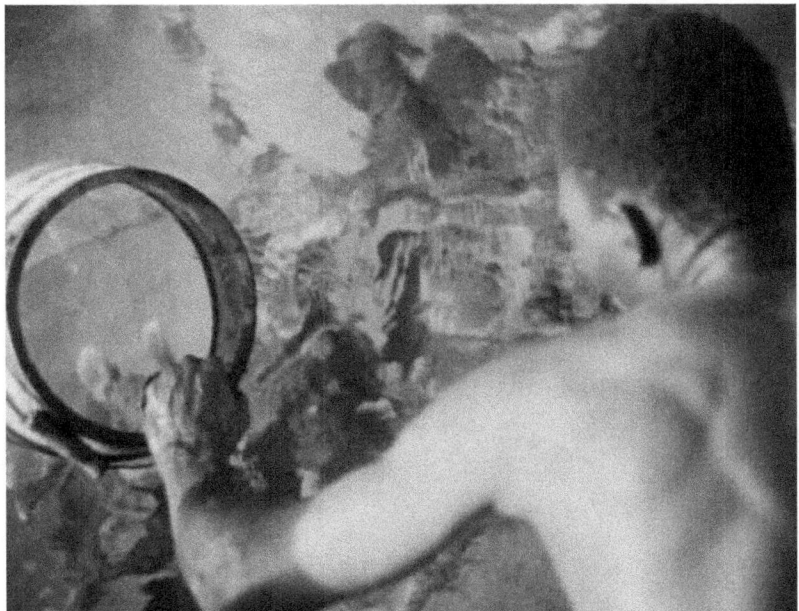

Fig. 5.2. Cylinder for magnifying the underwater. Film still. From J. E. Williamson, *Under the Sea*, 1929. Courtesy of the Field Museum Archives.

to remove dying coral polyps, as have tons of other specimens. Did they think the sea's resources were limitless? J. E. directs two men to run up the beach for no apparent reason other than to film the visual spectacle of African-Bahamian bodies as forms moving through space. The way Williamson has filmed the scene recalls Frantz Fanon writing on "the fact of blackness," in which he describes "being dissected under white eyes, the only real eyes."[26] Williamson's filming reduces blackness to a white idea of blackness, the body to an object, and exploits black athleticism as a commodity. The crates are loaded on a ship to show the operation is complete. Apart from a scene in which crabs run across the sand, the shots of nature are well and truly over, and the rest of the film turns mostly to human activities such as dockside work.

The second half of the last reel takes us under the sea to watch a diver with a knife attempt to kill a shark (fig 5.4). A stingray carcass is dangled in the water. The diver jokes as he faces the photosphere window brandishing the knife. Then a shark comes into view, cruising the sea floor, and takes a bite of the stingray carcass. Above the sea, members of the

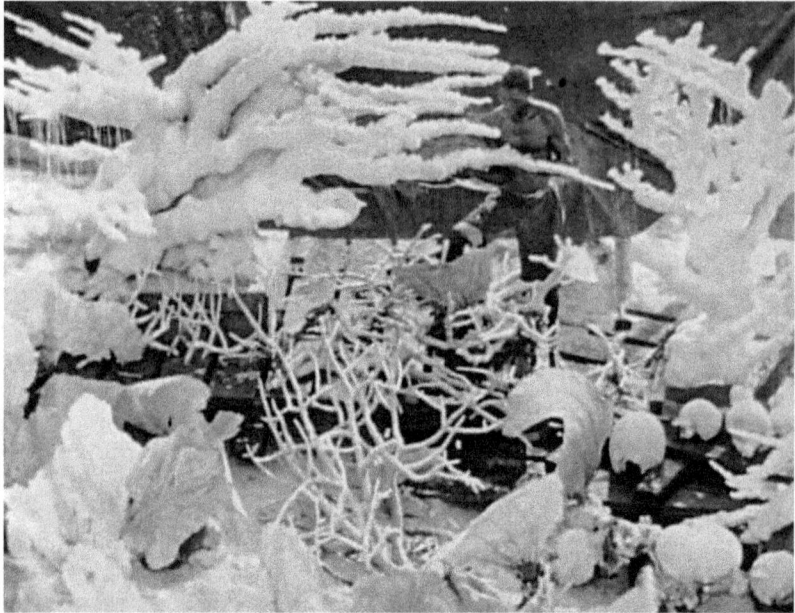

Fig. 5.3. Corals bleaching in the sun on the beach before transportation
to Chicago. Film still. From J. E. Williamson, *Under the Sea*, 1929.
Courtesy of the Field Museum Archives.

crew joke about hooking sharks and then lower a hook into the water to
catch one alive. Below the shark takes the bait. Next thing it is alongside
the boat. The film cuts to a scene in which the shark, a tiger shark, is attached to a rope in shallow water, only just alive. Three workers pull the
exhausted shark backward from the shallow water up the beach. They
punch it, and when it is almost dead they pour liquid down its throat until
it heaves and gasps (fig. 5.5). Then the camera shows Leon Pray with the
shark. He has already cut off its tail and fins and has molded the body in
preparation for the mannequin that it will become for the Bahama Islands
diorama. The jaws are bleached, and the molds packed and shipped. The
film returns to the map and shows the route the shipment will take back
to Chicago.

Jonathan Burt asks about nature films whether the medium of film
changes the way humans regard animals, or if audiences simply look at
and consume the vision of animals as imagery and object.[27] With *Under
the Sea* it is impossible to simply consume imagery. If anything, the vio-

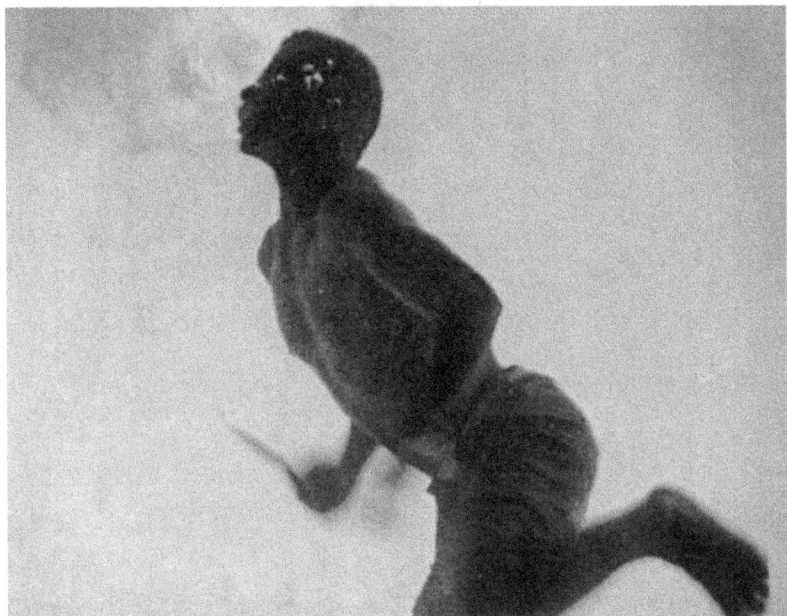

Fig. 5.4. Diver looking for sharks. Film still. From J. E. Williamson,
Under the Sea, 1929. Courtesy of the Field Museum Archives.

lence it exhibits to marine animals galvanizes the contemporary viewer
to form moral bonds with animals that appear in the film as sentient beings. *Under the Sea* is confronting for contemporary viewers who can em-
pathise with the lived experience of other beings, and are concerned with
the violation of the rights of others. They recognize something that Burt
is interested in—an "interaction of minds" as they look into the eyes of a
barely living shark and recognize a being that is thinking and feeling and
has an inner life.

Visually, the film uses the effect of light flickering across the underwater to mix beauty with foreboding. J. E. Williamson manipulated the
cutting of scenes by shifting abruptly from schools of pretty fish to lone
prowling sharks to ensure that the viewing experience was both mesmerizing and suspenseful. The divers add a menacing tone, too. Brandishing
knives, wearing heavy boots, at times hidden by fearsome helmets, sometimes grimacing underwater, they appear monstrous. In one final scene,
they take hold of a set of bleached shark jaws and mimic the biting actions

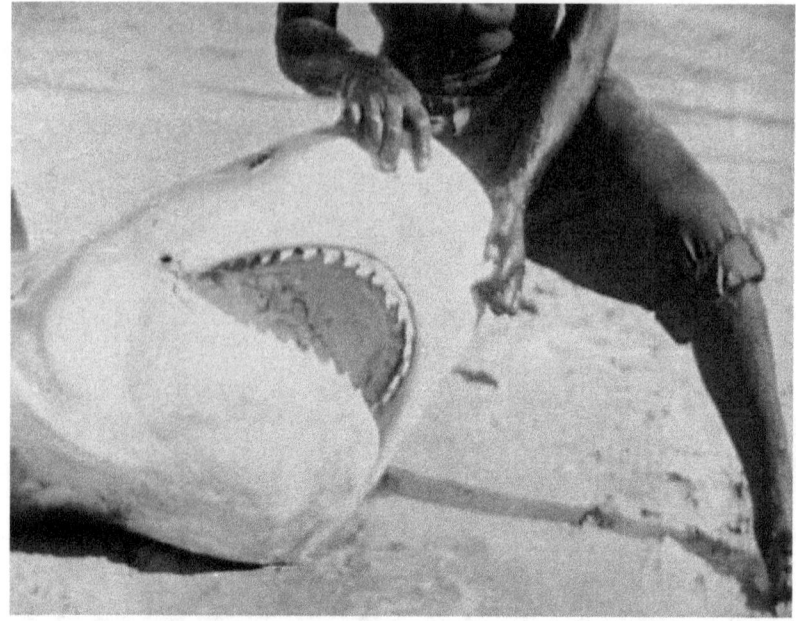

Fig. 5.5. Shark caught for molding. Film still. From J. E. Williamson, *Under the Sea*, 1929. Courtesy of the Field Museum Archives.

of sharks in a grotesque parody of animal savagery (fig. 5.6). This, coupled with the way they are so often seen on the other side of the photosphere's window, in and below the sea with sharks, fish, and corals, cements their portrayal as social outsiders.

The original purpose of *Under the Sea* was to create a natural history film and record a scientific expedition, but much of it involves hunting. Little wonder it found an interested audience at the Explorers Club in New York, where hunting and natural history were considered compatible endeavors, and where films of expeditions were regularly screened. Each time one of his new films was released, Williamson accompanied it on a lecture-screening tour, and in January 1930, at the conclusion of the Field Museum–Williamson Undersea Expedition, he presented a lecture screening of *Under the Sea* to four hundred members of the Explorers Club, during which the film stock caught fire.[28] Later, some of the footage of *Under the Sea* was spliced into a new film released in 1932 titled *With Williamson beneath the Sea*. When the later film was screened in New York, Mordaunt Hall, a journalist with the *New York Times*, remarked how "it is

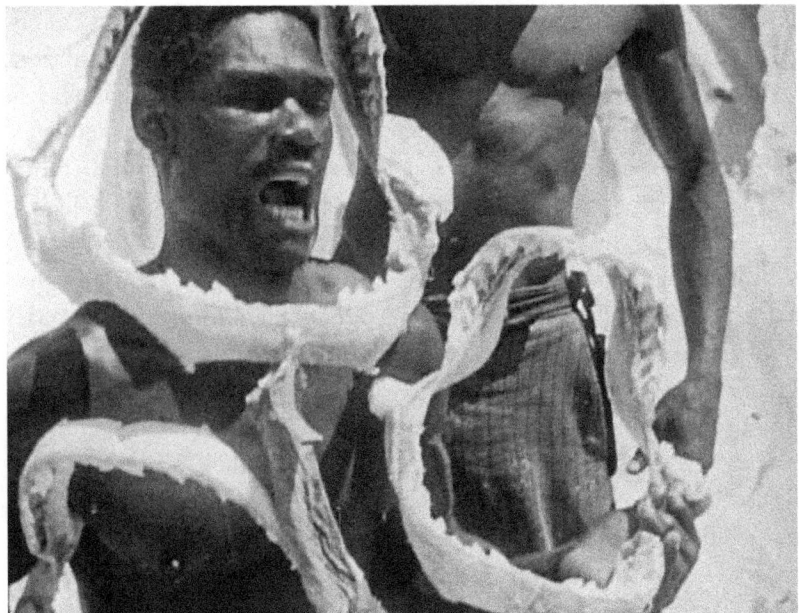

Fig. 5.6. Man imitating shark. Film still. From J. E. Williamson,
Under the Sea, 1929. Courtesy of the Field Museum Archives.

the type of picture that will be especially interesting to boys, for not only
are the divers seen under the water, with bubbles creeping up from their
helmets, and the barracuda and sharks, but there are also scenes of two
live sharks being hauled on the beach."[29] The gruesome hunting scene of
the tiger shark being pulled from the water had all the makings of a boy's
own adventure. The link between a boy's adventure and chasing animals
is discussed by John Miller in relation to early twentieth-century hunting
stories in which the chase of the animal is always the favorite part. He
argues that hunting literature, with its scenes of chase and violence, has
contributed significantly to the formation of male subjectivity.[30]

Williamson was an adventurer-hero of his own making and was in
many ways like the male heroes of nineteenth-century literature who
found themselves on coral islands beset by savages and sharks, or who,
in adventures such as Jules Verne's stories, were depicted as lonely figures surrounded by the sea. Diana Loxley, for example, notes that "the
Vernien world" is dominated by men, and that Verne's concept of frontier
space and "virgin territory is always-already controlled, its boundaries de-

lineated, by male reason."[31] It is a type of masculinity that explains the heroic nature of Williamson's enterprises. Modernity celebrated filmmakers who were also explorers and showmen and who blurred the boundaries between science and fiction. The popular press, for example, delighted in publishing stories about Williamson, the underwater, Williamson's films, and above all the photosphere as a magical story of heroism, technological modernity, and the ingenuity of engineering.

Ironically, given André Breton's gesture of acquiring Williamson's photograph and switching the geography of the Bahamas for the Great Barrier Reef, of the many international sites of reception for Williamson's films and photography, it was in Australia that they were fully embraced. From 1913 on, the myths and achievements of J. E. Williamson, and the machines that enabled him to photograph the bottom of the sea, were read about in every state of Australia through syndicated international news reports. At a time when the public had little knowledge of the vertical space of oceans, the very mention of "the underwater" suggested bottomless rather than shallow depths. The myth was that J. E. Williamson descended in the photosphere far below the surface, much further than the actual 250 feet that was the photosphere's maximum depth.[32] Consequently, more than once after 1916 newspaper articles reported on Williamson as a hero of the deep. One article in 1916 described his work as "the daring enterprise [that] illuminated the dark chambers of unfathomable waters round the islands of Bahama."[33]

Daring the enterprise was, and heroic as far as the public was concerned, but there was much exaggeration surrounding the nature of the work, the depths at which it was conducted, and the capabilities of the photosphere, a device, after all, that was in effect a moveable attachment to a ship, rather than an independent underwater navigational vessel. But the pleasure of viewing Williamson's films arose in large part from the thrill of imagining extreme depth, and from the image of a heroic adventurer. The Australian sense of manliness was wedded to explorers and frontiers, making Australia more than receptive to the image, and the imagery, of John Ernest Williamson.

CHAPTER 6

Williamson in Australia

Pictures of deep-sea life will soon be available as the
result of some unusual photographic work done in the
Bahamas. There is also a prospect that a new era is
about to open in submarine exploration.

The World's News, Sydney, Australia, 1914

The underwater as modern spectacle arrived in
Australia in 1914. It came with the release of J. E.
Williamson's films and with their promise of wilderness
and the deep sea, a new frontier for planetary exploration
in the underwater, and original photographic evidence.
The prospect of witnessing a previously invisible wilderness lay at the center of the excitement. But the trouble
with wilderness, as William Cronon explains, concerns
the misguided idea of an untouched, primitive region outside civilization. Instead, wilderness is a "profoundly human creation."[1] Frontier fantasies, though, give rise to heroes who also authenticate their masculinity in adventure.
So, while Williamson lived and worked among the beauty
of coral reef islands in the Bahamas, his public persona,
and aspirations for success, were founded on imagery suggestive of danger and struggle. His films therefore aimed
to immerse audiences in fearful fantasies, portray the racial Other as a natural part of primitive wilderness, stimu-
late imaginations with morbid fears of monsters, suggest
that the foreboding underwater harbors bitter struggles for

man and beast, and project the idea that because the deep sea is a wilderness, more fearful than the jungles of the land, it was a region that must be conquered.

Although, as Alain Corbin shows in *The Lure of the Sea* (1994), the cultural image of the sea changed from what he calls "the roots of fear and repulsion" in the eighteenth century to "admiration" by the middle of the nineteenth, it was more filmic and lucrative for Williamson to combine the two aesthetic principles of horror and pleasure in one.[2] In Australia, where his documentary and narrative films, photographs, and articles were eagerly received, Williamson was heroized as a great explorer of hidden worlds.

Expanding media networks and the rise of mainstream entertainment meant that even in a lonely outpost of the British Empire such as Australia, audiences could look forward to viewing Williamson's work within at least two years of publication, if not sooner. Attracted by the allure of the deep and its hidden creatures, they packed motion picture theatres in metropolitan and provincial regions. There was little delay, though, with press and magazine stories, many of which were written by Williamson himself and proclaimed him "a famous underwater explorer and photographer."[3] From these sources, Australians generated their own commentary on J. E. Williamson's achievements and the quality of his work, and among those who responded was a local hero known as James Francis (Frank) Hurley.[4]

How were Australian perceptions of oceans and ocean animals, and the tropics, shaped by Williamson's views and visions? What was the public impact of his films? The scope of public interest in Williamson's work can be summarized thus: adulation of pioneers and explorers; fascination with new technologies of exploration exemplified by the photosphere; fear of sharks, octopuses, and underwater monsters; and curiosity bordering on suspicion, anxiety, and voyeurism, from white people's perspective, about black people and racial difference. Williamson's films excelled at holding up a mirror to early twentieth-century confusion about oceans, ocean peoples, and ocean animals.

Bold self-promotion characterizes the articles Williamson wrote for the international media, particularly from 1914 to 1916, when the first films were released worldwide. In 1914, Williamson wrote how he had "accomplished the conquest of the deep" and given the public a visual passage to the frontier.[5] If tourists and travelers could not themselves visit underwater space, he would take them to the region he said was "unknown" and "undiscovered." In 1916, he claimed he was able to "descend deeper than any diver, and record for the screen submarine life and activity never before seen by the human eye."[6] There was no question in Williamson's mind that the underwater was uninhabited space—the agency of generations of maritime and Indigenous peoples who cared for it, and the undersea's inhabitation by animals and plants, did not figure in Williamson's Eurocentric and human-centered viewpoint, which was also the viewpoint of the vast majority of the audiences that bought tickets for his films.

Williamson hoped he would be seen as writing a new page in science, but his style of writing signaled showmanship and was more typical of popular, mass media entertainment than the comparatively reserved language of scientists. With his underwater filming, he planned to draw back the curtains on the greatest show on earth, revealing a primal scene beneath the sea among freak animals and amphibious humans. He would reveal another realm of existence, a world of nature that was outside and beyond the European world. When Williamson's early films of the underwater were first released in 1914—films that were made when he was still in partnership with George—they brought to the lives of everyday people, who had never laid eyes on the undersea, a mixture of magic and terror. Ghostly "forests" of corals, strange fishes and sea creatures, and, of course, the "menace" of sharks infused Williamson's natural history films with suspenseful drama and narrative. His first films were silent and black-and-white, but in those early days of public cinema the sight of the underwater as a living, moving phenomenon brought to life many aspects of nature that had previously only been imagined through illustrations, myths, and books, especially stories and illustrations in books by Jules Verne: "The marvels that Jules Verne saw through the prehensile eyes of his vivid, Gallic imagination, the Williamson brothers are now seeing through the eye of a camera at the bottom of the sea."[7]

Williamson made an impression on Australia with the release of *20,000 Leagues under the Sea* because, as one reviewer noted, it appealed to a generation raised on literature and took audiences back "to the days when we reveled in the marvelous imagination that Jules Verne displayed in 'Twenty Thousand Leagues Under the Sea.'"[8] As well, the underwater scenes were described as a revelation that permitted spectators "to see the actual bed of the ocean with the marvellous marine gardens and coral formations."[9] As late as 1940, Williamson was known in Australia as "a modern Neptune."[10] But if, in 1940, he was seen as something of a god of the underwater, rapid changes in the science, technology, and representation of oceans throughout the twentieth century eventually overshadowed Williamson's influence on the Australian imaginary. Today he is largely forgotten, despite giving many Australians who were the audiences for his films between 1916 and 1940 their first views of underwater ocean life.

Williamson's "wildlife," "natural history," and "documentary" works had quite an effect on their audiences. With the exception of the documentaries of African animals by Cherry Kearton (1871–1940), which Kearton filmed and released between 1910 and 1913, wildlife films were relatively new to audiences.[11] And the wildlife films that audiences did see were filmed on land, including underwater scenes that were shot through the glass sides of aquariums. The fact that Williamson's films were captured underwater distinguished him and made his documentary work a novelty.[12] One reporter described how Williamson had broken "the boundary between the world of men and the fantastic realm of colorful coral and equally colorful fishes."[13] Yet because there was so much intervention of technology in the production of Williamson's wildlife films—which were not only shot underwater from the encasement of the photosphere but were also likely produced using telephoto lenses that gave viewers the illusion of being close to subjects, and lit with artificial lights that affected the behaviors of wild animals—they were closer to popular entertainment than to documentary, a distinction that Derek Bousé makes in relation to early wildlife films that used similar techniques.[14]

Nevertheless, in the earliest days of public cinema, when film had a magical aura, the idea that a filmmaker could also take an audience to the unseen and unknown underwater realm to view marine animals in the wild was almost beyond comprehension. By 1913, news of Williamson's explorations in the photosphere had reached Australia through the popular press, which continued to publish stories of interest about Williamson well before the first movies were seen in 1916.[15] The photosphere came to be known in Australia as a "camera barge," symbolizing a new phase of camera technology that was designed for depicting oceans.[16] That Williamson and his ideas were not limited by convention and not grounded in reality brought further mystique to his enterprise. He was seen as daring and modern, while other photographers and explorers were seen as traditional and safe. As stories were published and republished through newspaper franchises in every state and territory of Australia, Williamson insisted that the invention of the photosphere was as revolutionary to undersea exploration as the modern ocean liner was to sea travel compared to a "dugout canoe."[17] His pictures, he claimed in the same article, were a vital new page in natural history. His project to photograph the bottom of the sea, he said, denoted "perhaps the last of the big things to be done on this hoary planet."[18]

By 1924, the reception of Williamson's films in Australia began to reflect a palpable ambivalence toward the sea, and especially toward marine animals. While on the one hand, a review would celebrate the "wonderful sea pictures," on the other it recognized how frightening nature could be in coral reef environments in which "startling beauty" stood in marked contrast to "the grim struggle for life which is forever going on about them."[19] Through the media, Williamson's stories and images were imprinted on the Australian psyche and his name associated with the terrors of coral reefs. Lurking behind, under, and between the structures of growing corals was the otherness of strange and predatory animals. Williamson's films and photographs, as well as the articles he penned for publication in Australia, had significant potential to turn people against marine nature. One Australian journalist, Ewen K. Patterson, agreed that the dangers that Williamson said plagued Bahamian reefs underwater were no different than those of the Great Barrier Reef of Australia: "It is true that beneath the beauty of Nature, everywhere, horror lurks. Beauty and horror; good and evil—the opposites."[20]

Patterson's list of concerning creatures included octopuses, giant clams, "pugnacious eels," poisonous stonefish, and sharks. The fear of octopuses was incited by articles that Williamson himself penned and published in Australia. Like sharks, octopuses symbolized for Williamson everything that was unpredictable and evil about the sea: "No words can adequately describe the sickening terror one feels when from some dark mysterious lair the great lidless eyes of the octopus stare at one. People speak of the cold eyes of fishes, of the cruel, baleful eyes of sharks, but in all creation there are no eyes like those of the octopus. They are everything that is horrible. Dead eyes. The eyes of a corpse through which the demon peers forth, unearthly, expressionless, yet filled with such bestial malignancy that one's very soul seems to shrink beneath their gaze, and cold perspiration beads the brow."[21] Moving images, though, were more powerful than the written word for influencing public opinion, and no audience could escape the emphasis that Williamson placed in his films on portraying the terror of sharks, nor the way he compounded the horror by involving the animal in struggles with "native divers." His films and writing featuring sharks typified a prevalent modern stance toward animals, a standpoint that Jonathan Burt explains in relation to modern cinema as "the unresolvable dialectic between humane and cruel attitudes."[22]

"Mr Williamson," reports one Australian newspaper in 1931, "regards all sharks as man-eaters."[23] The report was based on an article written earlier the same month by Williamson, in which he stated, "And is he a maneater? I say 'yes.' I treat them all as maneaters. While one or two species lack the bristling rows of teeth that most of them carry, I distrust them all, for I know from many years of experience with them that when they have once had a taste of blood or flesh, they will attack like a mad dog, and one cannot tell when the killing mood may seize them."[24]

Australian newspapers published horror stories about sharks in the Bahamas, and it was common to read that the wider Caribbean area was "shark-infested." The Bahamas and sharks became synonymous, although the history of this in Australia went back to the late nineteenth century, when in news about fishing in the Bahamas it was said: "We never let an opportunity pass to kill a shark. No one ever does. The waters [are] full of them."[25] And so it went on for decades, peaking in the 1920s, when Williamson's influence was most intense, and reaching its most irrational conclusion in 1923, when a report from Nassau claimed, "The shark is against everyone, everyone against the shark."[26]

Eventually, with so much hunting and so much hatred, and the unresolvable dialectic between humane and cruel attitudes, the shark populations of the Bahamas, and of Australia, became markedly depleted.[27] Sharks at reefs were killed to near extinction.[28] An issue of one of Australia's most popular magazines, The Home, from 1928, offers insight into the depth of shark hatred in Australia in the early twentieth century. In a story about exploring the Great Barrier Reef titled "Meet Us at the Reef," the pages are illustrated with photographs of dead sharks strangled by ropes and choked by hooks. It is a sight familiar to many Australians over the age of fifty, including the author, Tim Winton, who describes the once common practice of hanging sharks by the mouth as "public executions, the criminal species strung up for the crowds, as if the only good shark were a dead shark and we needed to see this butchery acted out again and again for our own wellbeing"[29] (fig. 6.1).[30]

On the subject of Williamson's impact, particularly on attitudes to sharks, it is difficult to overestimate the sensation caused in 1916 when

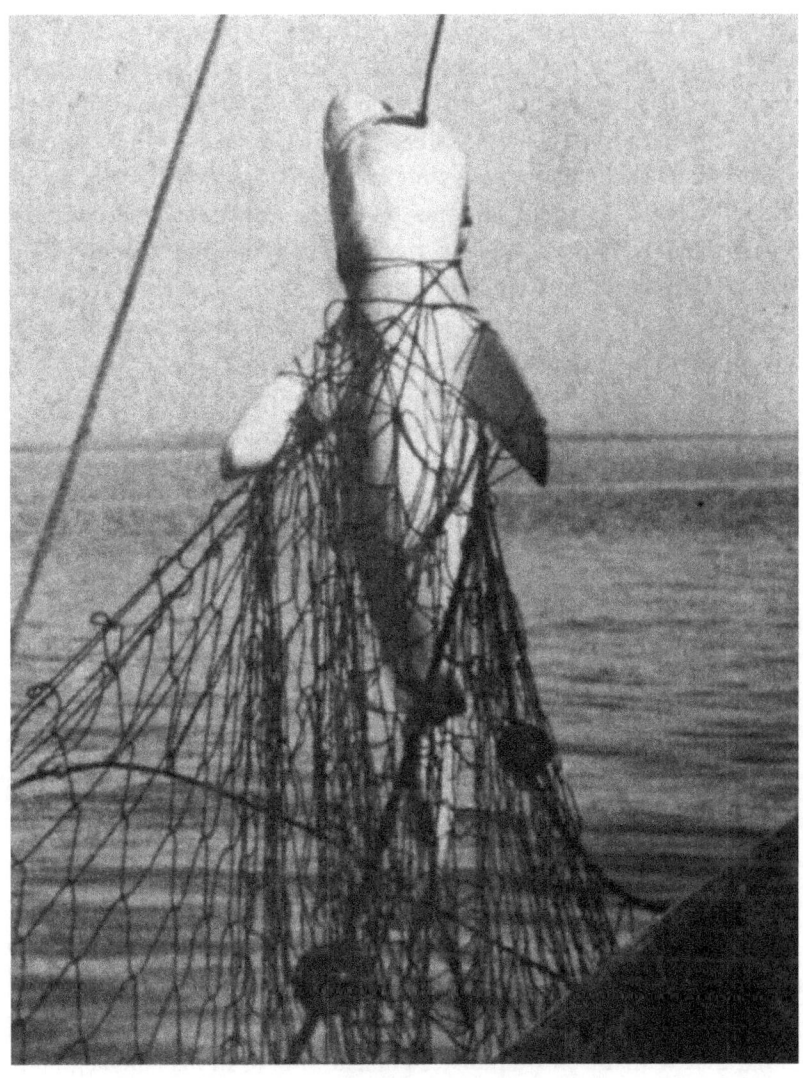

Fig. 6.1. "The Shark Ascending: Meet Us at the Barrier," *Home*, May 1928, 44. Photograph by T. C. Roughley.

The Williamson Submarine Expedition was first screened in Australia.[31] The film was made in 1914 but was screened in Australia in 1916, when it was acclaimed as "the most wonderful thing yet accomplished in moving pictures."[32] It was, according to Thomas Burgess, "one of the first feature-length nature films ever put together" and therefore promised to give audiences a completely new perspective on the ocean.[33] The film had two draw cards: it was "the only genuine motion picture ever taken at the bottom of the sea," and it showed "a deadly combat between a diver and a monstrous shark."[34] The first Williamson film to screen in Australia, then, was the one in which he famously killed a shark with his bare hands underwater. With the morbid fascination that Australians already had for sharks, Williamson's stories and films, witnessed from the safety of the picture theatre, found a receptive and enthusiastic public.

Screened in 1916, right in the middle of the First World War, the scene in which Williamson fights and kills a shark reinforced the symbolisms that were foremost in peoples' minds: good versus evil, foe versus foe, survival and death. One reviewer noticed that when the shark was "in its death agony," the water became suffused with blood but despite the gruesome sight recommended the experience of the film to others because "it is the sort of thing which must be seen if one would thoroughly realize the deadliness and dangers of the conflict."[35] Moreover, Williamson's film was double billed with the British film *With the Fighting Forces of Europe* (Charles Urban, 1914). With conflict and danger on everyone's mind, it was easy enough for audiences in Australia in 1916 to militarize the underwater and transpose on the shark a symbol of the enemy, Central Powers.

Two years before the film was released in Australia, an image was published in an article in the *World's News* advertising *The Williamson Submarine Expedition* and promoting the kinds of exciting underwater imagery that audiences could soon expect to see when the film finally arrived in movie theatres (fig. 6.2). The image was an illustration based on a photograph that Williamson would later publish in 1936 in his autobiography *Twenty Years under the Sea* (fig. 6.3). It is also very similar in style to the silhouette of a black diver wielding a knife that Williamson found filmic when shooting *Under the Sea* in 1929 (see fig. 5.4).

However, the illustration in figure 6.2 is not true to *The Williamson Submarine Expedition* because it does not show Williamson, a white man, killing a shark; instead it shows a "native diver." During filming, a scene was in fact shot of a Bahamian diver killing a shark with a knife—but

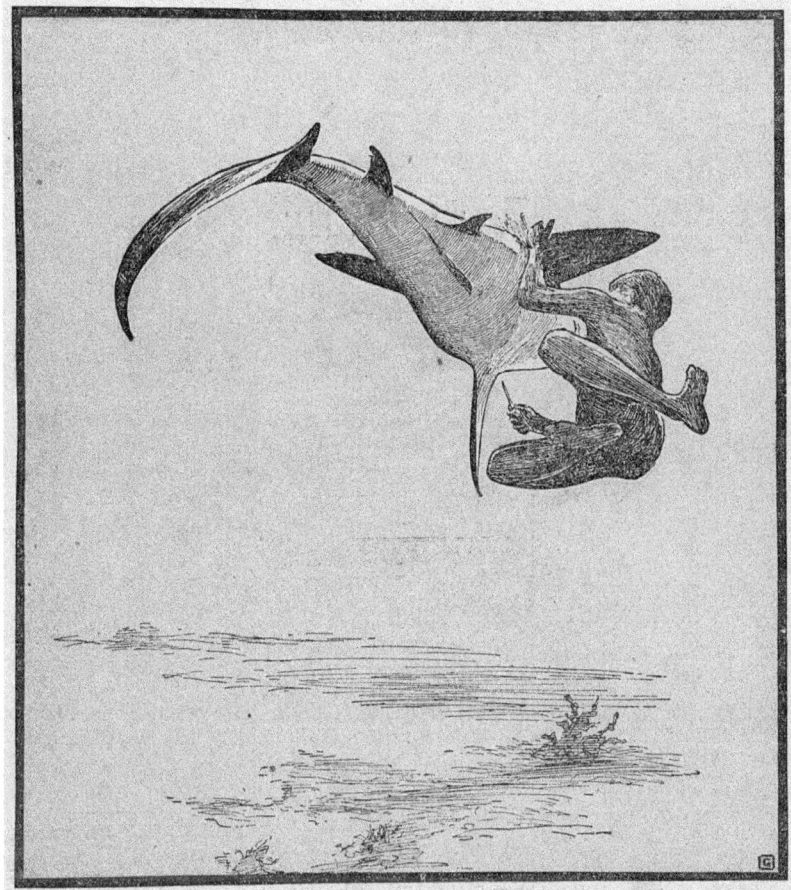

Fig. 6.2. "Fight between Man and Shark," *World's News* (Sydney),
September 5, 1914, 1 – 2.

the action wasn't caught on film because it took place outside the viewing
space defined by the photosphere's window.[36] Williamson himself then
stepped in and fought the shark.

Figure 6.2 shows the diver from below about to knife a shark in the
belly. A comparison with the photograph it was based on, seen in figure
6.3, reveals changes to the viewing angle as well as the inclusion of a reference to the sea floor to help orient the viewer spatially. The photograph in
figure 6.3 that the illustration is based on is meant to look like a film still
but is in fact a publicity image. The shark is limp, not fighting; it is clearly
dead, and the photograph a setup to simulate an attack. Without fast film,

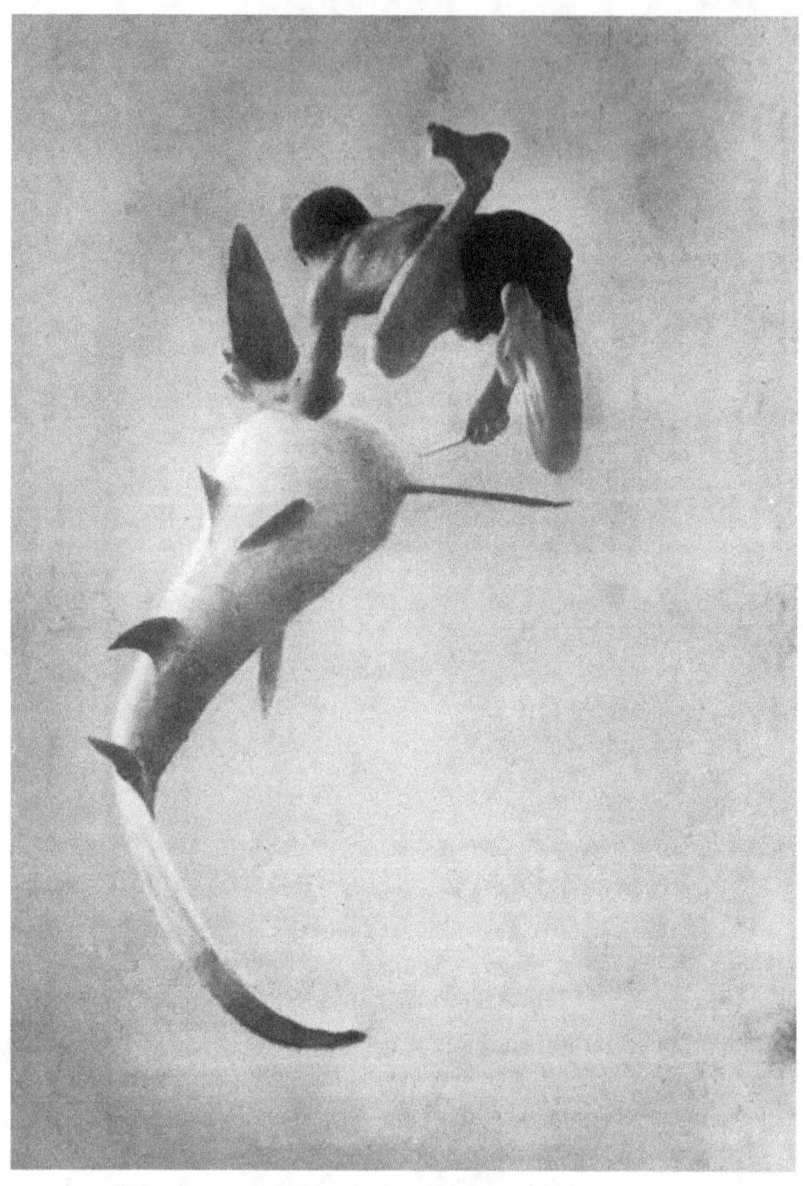

Fig. 6.3. "The diver drove his knife up to the hilt into his enemy."
From J. E. Williamson, *Twenty Years under the Sea*, 1936, 68.

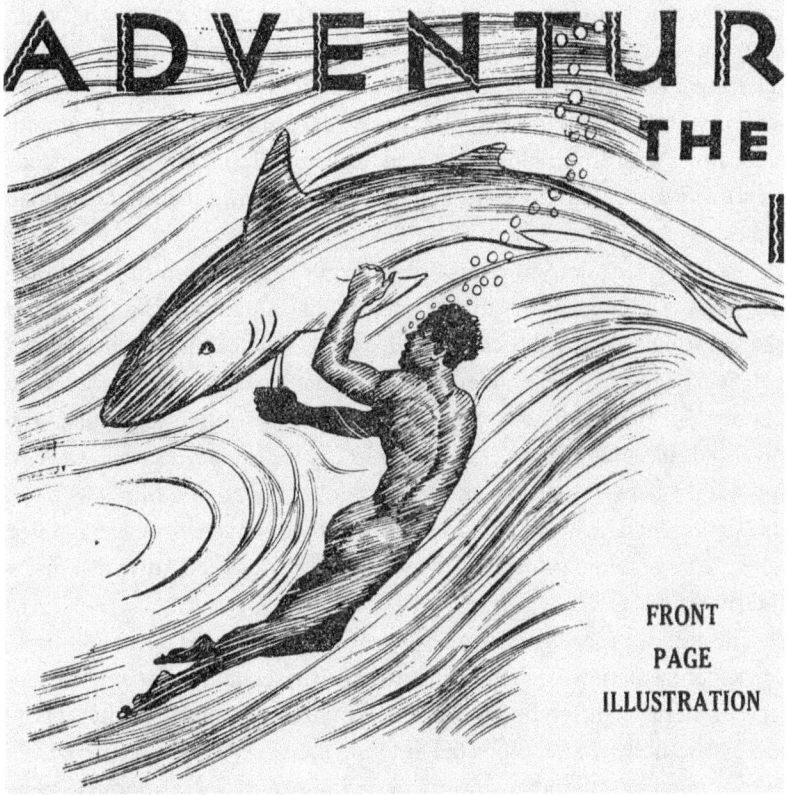

Fig. 6.4. "Adventures on the Ocean Floor," *World's News (Sydney)*,
May 27, 1931, 16–18.

it was impossible in 1914 to capture a shark streaking through the sea and obtain a clear image.

The inclusion of a "native diver" in the publicity image for *The Williamson Submarine Expedition* added greater excitement. The idea was later adapted in slightly different form to advertise another of Williamson films, *Adventures on the Ocean Floor* (1931; fig. 6.4). Both illustrations—of a "native diver" fighting a shark published in 1914 to promote *The Williamson Submarine Expedition,* and the scene of a fight between a Bahamian diver and a shark in *Adventures on the Ocean Floor*—represent the racial Other as an exotic, primitive object of nature. And film reviews, including commentary about *Adventures on the Ocean Floor,* do the same: "The ebony body of a full-blooded African negro, naked but for a loin-cloth, poised a second on the sloop's rail; then, gleaming in the sun, shot through the air, cleaving the crystalline water without a splash. Down he struck through the limpid sea, darkening as he swam out of the sun's range."[37] Exaggerated are the bodily qualities of blackness, nakedness, wetness, and propulsion, along with amphibiousness, and the symbolic darkening of the body as it disappears to the ocean's depths to the shadowy realm of untamed forces.

The experience of watching Williamson's films went well beyond being passive witness of the underwater, including animals and events; it shaped and determined social attitudes to the sea, to animals, and to people. Social "truths" were taken from the constructed and fictionalized experience of the media spectacle. In fact one Australian journalist who reviewed *Adventures on the Ocean Floor* in 1931 rightly argued that witnessing a fight between a diver and a shark in a Williamson film was more like fiction than fact, and noted a scene in which suspense was heightened when "a large negro propelled by swift, strong strokes . . . turned and saw the shark."[38]

Through film screenings, newspaper illustrations, and books, Williamson's representations of racialized difference in relation to the sea, his projections of racial difference onto animals of the sea—and of animals of the sea, particularly sharks projected onto the racialized Other—were widely encountered by Australians over a period of twenty years and need to be considered in relation to how race-thinking was shaped in Australia. In 1922, for example, at Coogee, a beach in Sydney once described as "shark infested," eighty thousand white Australians gathered to watch nine Loyalty Islanders of the South Pacific take to the surf in loincloths to hunt and kill sharks with knives. Spectators were disappointed when no sharks were sighted, but the story shows that by 1920, the spectacle of Indigenous men fighting sharks had become a way of defining Islander peoples, and something of a blood sport in Australia.[39]

The motif of shark and "native diver" became a common sight in Australian popular culture illustrations of the twentieth century. In 1948, for instance, Sanitarium, a food company that produced breakfast cereal, issued a set of pictorial cards of the Great Barrier Reef for children to collect. One card shows a naked diver at the Great Barrier Reef, a figure not distinguished by name as Aboriginal or Torres Strait Islander but implied to be Indigenous. The male figure is shown striking through the underwater followed by a shark, both man and animal transfixed by the urgency of chase and escape (plate 5). Also in the card set was a variation on the shark-versus-man theme: a struggle between an Indigenous diver and a giant clam. The viewer is left to imagine the man drowning or being eaten alive, if not by the clam then by a shark (plate 6). From a very young age, then, Australian children learned through popular imagery to recognize and stereotype racial difference and species difference, to imagine both the racial and animal Other as savage and terrifying, and to displace animality onto the bodies of Indigenous peoples. It was part of a visual culture that emerged in the late nineteenth century and that Paul Gilroy describes as "celebrating and creating a stimulating world of signs to which racial difference was absolutely fundamental."[40]

Chilling but false stories circulated in the press warning swimmers that the giant clam "floats about the water, always open, and woe betide any poor unfortunate who steps into it!"[41] Many stories about coral reefs were mythical, and they were often frightening. In the early twentieth century, the image people had of coral reefs was not always the picturesque retreat that occupied the imagination after the expansion of tourism in the 1950s. The thought of a living, tangled edifice of limestone grottoes, of plants and algae, animals, rocks, and unknown monsters lurking beneath the surface of the sea meant the coral reefs of the Great Barrier of Australia in the early twentieth century were portrayed in the press as both a fearful and inhospitable force of nature, as well as a beautiful, peaceful environment of spectacular colors and forms.

The Australian press covered most of Williamson's adventures, including the expedition to Scotland to hunt the Loch Ness Monster. His plan to descend into Loch Ness to determine whether the monster was a sea serpent and film the outcome seemed exciting.[42] All the more so since Australia had a monster mystery of its own. A creature resembling a giant sea serpent had been sighted at Fraser Island and at other sites at the Great Barrier Reef. In 1891, Selina Lovell, a schoolteacher on Fraser Island, reported seeing the monster in shallow water, and even made a sketch of the creature, which William Saville-Kent later reproduced in his book *The Great Barrier Reef of Australia* (1893; fig. 6.5).[43]

In 1934, the same year Williamson set off to find the Loch Ness Monster, a party of scientists from Melbourne arrived in Townsville in northern Queensland to hunt the monster of the Great Barrier Reef. The party's leader believed the "monster was a giant water tortoise."[44] Fishermen on the outer reef had seen what they described as a sinister and giant snake

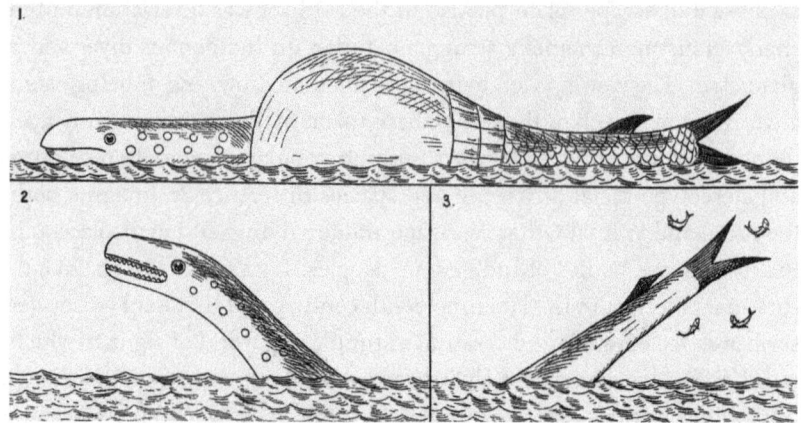

Fig. 6.5. "The Moha-Moha, or Great Barrier Reef Sea-Serpent." From William Saville-Kent, *Great Barrier Reef of Australia*, 1893, 324.

towering out of the water. One report in Brisbane's daily tabloid newspaper, the *Courier-Mail*, said its "feet were like an alligator's; the head and long neck moved under the great dome-shaped carapace; the tail was bifurcated and fleshy; a fin was thick and fleshy."[45] Aborigines knew the creature and had named it Moha Moha. Was the Great Barrier Reef, with its immense, unexplored reef structures, the home of sea monsters?

Invite J. E. Williamson to Australia, urged one journalist in 1935. Get him to bring the photosphere and explore the Great Barrier Reef underwater. Neville de Lacey felt Australians were missing out on adventures and explorations of their own simply because there was no national effort to make the region's oceans and reefs, especially the underwater, known internationally. In a local newspaper, he wrote, "I have often wondered why no effort has been made to induce Mr. Williamson to bring his photosphere (as his under-sea cinema apparatus is called) to the Great Barrier Reef. A live Government (if we had one) could easily interest this enthusiast in the most beautiful sea-bed probably in the world. People in other portions of the globe know very little about the Barrier, for the reason that Queensland never tells them anything."[46]

As history shows, although Williamson did transport the photosphere to locations outside the Bahamas, such as Scotland and California, the photosphere was never brought to Australia. This was no doubt a source of disappointment to the machine-obsessed culture of the early twentieth century, to whom the photosphere was as much a celebrity as Williamson.

The photosphere elevated Williamson's status as an explorer compared over that of other would-be explorers of the underwater whose ambition in the 1920s was also to photograph with cameras submerged in the sea. One Australian explorer, though, who thought Williamson's photosphere had limitations for undersea work, was Frank Hurley.

Frank Hurley, an explorer, photographer, and filmmaker of renown, claimed in 1926 to have greater mobility in the underwater than Williamson. He claimed he used diving gear to submerge his body in the underwater, had waterproof cine-cameras, and wasn't encumbered by the clumsiness of an encasement such as that of the photosphere. In 1920, Hurley publicized his ambitions to photograph and film the "floor of the sea," and to do so among the coral reefs that stretch from the east coast of Australia to Papua New Guinea. No one had yet photographed from beneath the surface among the coral reefs known collectively as the Great Barrier Reef.

It was in 1926, in an article for the Australian newspaper, the *Sun*, that Hurley compared himself to Williamson in such a way as to make himself appear more advanced technologically and more daring in his ambition for underwater work. In 1926, Hurley was at the peak of his career. In the same decade, he had completed two expeditions to islands of the Torres Strait and Papua. He had exhibited worldwide a remarkable travelogue of the Torres Strait and Papua titled *Pearls and Savages*. The film was first released in 1921, and was amended in 1923 for markets in the United States and Britain. In 1926, Hurley was funded by the Australian-born British film magnate Sir Oswald Stoll (1866–1942) to make two feature-length narrative movies—*Jungle Woman* filmed in Dutch New Guinea, and *The Hound of the Deep*, later released in Britain as *Pearl of the South Seas*, a movie filmed on Thursday Island in the Torres Strait that included scenes of pearl divers underwater.

As a prelude to discussing the underwater sequences of *The Hound of the Deep*, Hurley gave readers of the *Sun* in 1926 a brief history of underwater photography, and into this history he inserted himself and Williamson, but in ways that both affirmed Williamson's place, and slightly trivialized it. By 1926, it was J. E., not George, who was an international media celebrity, but Hurley referred to "the Williamson brothers" as pioneers of underwater exploration and filmmaking:

Underwater kinematography is an expensive and uncertain operation in the most favorable circumstances. The Williamson brothers—world-famous for this work—use a caisson—an airtight chamber fitted with glass ports. This can be entered from the surface by way of a steel tube, which is secured to a barge or other suitable vessel. When employing this apparatus it is essential that the sea be calm, as any movement of the vessel is greatly amplified in the chamber below.

Sometime ago I made satisfactory experiments in Sydney harbor with a watertight camera operated electrically. I made many descents with this equipment, and as a tribute to its efficacy, the underwater scenes in "Pearls and Savages" were made with it. Though less certain of definite results than the Williamsons' outfit, my equipment has the advantage of greater mobility: further, it can be used under any sea conditions when diving is possible. I am using this apparatus for taking the deep-sea scenes depicting divers gleaning their harvest of pearl shell from the depths of the Coral Sea in my new picture. The more intimate close-up shots portraying the travails and dangers of the deep have already been done in a huge aquarium.[47]

Was Hurley more mobile in the underwater than Williamson, and could he dive and operate watertight cameras immersed in the depths of the Torres Strait? Mobility was a sign of an explorer's modernity, and, as Arthur B. Evans explains, "unlimited mobility" was a utopian ideal inspired by Jules Verne's stories of modern explorers taking charge of every dimension of the planet.[48] Mobility was a sign of competitiveness, but for Hurley to draw attention to his own greater mobility as compared to Williamson's was also a case of one-upmanship. It was a competition over whose approach to underwater photography had more freedom, and who was more agile in the water.

There is a great deal to unravel in the passage just quoted from Hurley's *Sun* article, but the points that are conspicuous and that need to be untangled form the basis of chapters that follow: Hurley's claim about having greater mobility than Williamson; the significance of diving to Hurley's self-image; the nature of underwater scenes in *Pearls and Savages*; deep-sea scenes of pearl divers in *The Hound of the Deep* (*Pearl of the South Seas*); and Hurley's reference to aquariums.

At this point, it is wise to keep in mind what Andrew Pike and others have claimed about Hurley: that he "fabricated exciting stories about himself for the public."[49] And in this regard, he shared much in common with J. E. Williamson. In 1920, Frank Hurley embarked upon the first of two

filmmaking expeditions to the tropics north of Australia. He said he possessed the newest, most modern cameras and gadgets, "equipment," he
told the papers in 1920, that "records in natural color [and] is adapted for
aerial, surface, and submarine photography."[50] He said the expedition was
for science, and he promised the public that he would return with films
of the deep at the Great Barrier Reef.[51] Just how deep he intended to go
is unclear, but the very mention of the word "deep" suggested something
profound, heroic, and challenging.

About the impending expedition along the Great Barrier Reef to the
Australian-administered region of Papua, one newspaper expressed awe,
particularly at the scope of the project: "To the marvels of the coral built
tropics and the mysteries of unknown Papua Captain Frank Hurley the
famous Antarctic photographer has gone for new copy and studies. On
the sea, beneath the sea, on the earth and above the earth he will be busy
for some months."[52]

Above the world and below, Hurley claimed to have the capacity to take
charge of any viewpoint. In the years following the First World War, when
the skies and the deep became militarized spaces, he imagined his body
in relation to a vertical space that was no longer simply terrestrial. What
the tropics to the north of Australia offered, especially the underwater at
the Great Barrier Reef, was something that Robert Dixon has called the
modern commodity of the "primitive" geographies and peoples of the colonial tropics.[53] David Millar, Hurley's biographer, put Hurley's ambition
in perspective when he commented about the social context for Hurley: it
was "a period in time which has now gone forever, a period of empire, of
high endeavor, of men doing great things."[54]

Turning to Frank Hurley at this point puts J. E. Williamson's story on
temporary hold, but Williamson rejoins the narrative in due course when
both men are examined in the context of explorer culture. The reason to
turn to Hurley is to investigate the claim that in the field of underwater
photography and filmmaking he had achieved what J. E. Williamson had
not: mobility in the underwater by diving with waterproof cameras to record the spectacle of the coral reef.

PART III

*Frank Hurley and
the Great Barrier Reef*

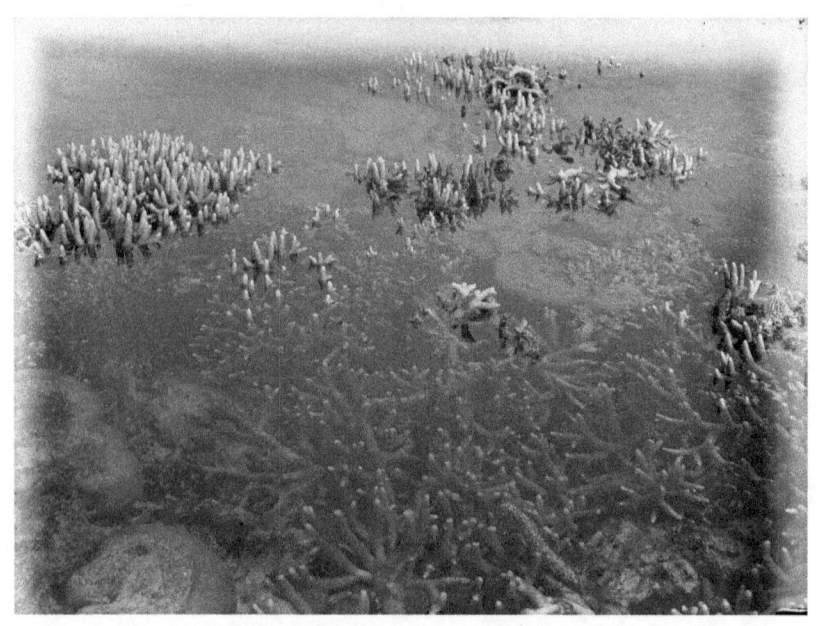

Fig. 7.1. Frank Hurley, "Exposed coral," Torres Strait, 1921.
National Library of Australia, NLA PIC FH/8913.

CHAPTER 7

Hurley and the Floor of the Sea

The history of descending into the depths of the ocean is filled
with dubious facts, half-truths, and improbable claims.

Firebrace, "Aquarius in Question"

If it is the explorer's task to discover unknown territories and test their own personal limits, what did
Frank Hurley hope to discover in the underwater on the
two expeditions he made to the coral reefs of the Torres
Strait, Papua, and New Guinea in 1920 and 1922? When he
left Sydney in 1920, headlines pointed to all the sensational
frontiers he planned to conquer: "the floor of the sea," "the
underwater," "unknown Papua," "a prehistoric wilderness,"
"the wilds," "the jungle," "lost worlds," "[life] beneath the
waves," "darkest Papua," "spirits," and "the wonders of the
deep."[1] The type of experience we imagine he sought is
characterized by David Millar as the stoicism of "men challenging their environments, proving their moral worth and
earning the public adulation due to all heroes."[2] The concept of "the deep" had rhetorical power, but what actually
captured Hurley's imagination when he reached the tropics was the tamer, more intimate, less fearful, and more
beautiful but still marginal underwater space belonging
to the relatively shallow depths of the reefs.

What seized public interest about Hurley's expedition
plans in 1920 was the impressive mobility implied by the
scope of the journey, Hurley's competitiveness, and the

risks he was prepared to take on an adventure that would take him through Cape York Peninsula to the Northern Barrier Reef and then on to British New Guinea. The public was surprised to learn that Hurley intended to take with him "a complete diving and underwater apparatus," with which, he said, "I hope to secure films of secrets of the deep submarine life along the barrier."[3]

The same information is recorded in the first diary entries: "My outfit is of the most complete. Two cinematographs one adapted for underwater. Also two still cameras & a vast quantity of material for every contingency."[4] As he sailed north to the Torres Strait, the thought of succeeding at filming and photographing the tropical undersea preoccupied Hurley's mind. He knew it would send the public a strong message about his status as a modern hero. Australian papers were full of news about the underwater frontier. Not only were there many reports about J. E. Williamson in the Bahamas, but considerable coverage was also given to a British-American artist, "Zarh" (Walter) Pritchard (1866–1956), known as the man who painted corals, fish, and seaweed underwater with oily crayons while wearing a diving rig.[5] Zarh Pritchard insisted that the true, natural qualities and features of the underwater could not be represented "by any method of observation from the surface."[6] Williamson and Pritchard, then, who both brought back visual images from the underwater, set important benchmarks for their generation in undersea representation.

In the 1920s there was much mystique in Australia surrounding the futuristic and modern subjects of diving and underwater photography. Helen Rozwadowski points out how the submarine realm at that time was understood as a challenging place to "work, play, fight, explore, and reflect at, and under."[7] Intrigue with diving, combined with a public appetite for stories and news about islands inhabited by "cannibals," of expeditions through jungles, about seas sparkling with coral reefs, and of the unseen underwater, led the *Sun* newspaper to buy exclusive rights to publish Hurley's expedition accounts and photographs. Through telegraphy, mass print distribution, and popular newspapers Hurley would supply readers with an escape from everyday life to exotic places that were named in the press as "stone-age Papua" and the "primordial underwater." The idea of escaping there through the spectacle of cinema and photography further strengthened the binary through which Western society defined itself, the dualism of the modern and the primitive.

Hurley's ambition was to photographically colonize as much geographical space as possible, and supply a business demand for films and photographs of exotic domains of the planet. In 1920, when it was claimed that exploration of "the sea depths has hardly begun," the sea offered more exciting potential than the land.[8] A bustling maritime culture of Indigenous, itinerant, and colonial workers defined the Torres Strait, Papua, and New Guinea in the 1920s. At that point in time, Papua was the southern half of eastern New Guinea and included outlying islands. It was an area colonized by Britain as British New Guinea but from 1902 was administered by the Australian Commonwealth as the Territory of Papua. The northeastern part of the island, which was colonized by Germany and known as German New Guinea, became a mandated territory of Australia in 1920. Western New Guinea, now part of Indonesia, was also known in Hurley's time as Dutch New Guinea. Australia administered the combined regions in the east until Papua New Guinea's Independence in 1975. But when Hurley ventured forth, the Torres Strait Islands, Papua, and the former British New Guinea were known as Australia's "tropical possessions" and "our unknown lands."[9]

In the context of the times, Hurley's idea of getting under the surface of the sea and revealing it through moving and still pictures involved an expansion of vision that paired naturally with the enterprise of colonial expansion. The first expedition was funded by, and conducted on behalf of, the Anglican Board of Missions, and Hurley's job was to make films and photographs of its work while the board, in return, would assist Hurley as he traveled from island to island meeting with missionaries, administrators, and planters, and observing and recording the lives and environments of Indigenous peoples. Hurley was rather like the figure of the colonial wanderer in Joseph Conrad's *Heart of Darkness*, a book he knew.[10] The colonial wanderer in Conrad's book was described by Edward Said as one who wanders "colonial regions, telling his story to a group of British listeners," primarily from the business world, turning "the business of empire" into "the empire of business."[11] The second expedition, in 1922, was privately funded, largely by the proceeds of *Pearls and Savages*, the film that Hurley completed after the first journey.

During the first expedition, Frank Hurley dispatched dozens of news stories to the *Sun*. They were written in a formulaic and melodramatic style that mimicked archetypal tales of the unknown. What drew him to the underwater, he said, was the promise of intense adventure and uncanny encounters with strange and alien creatures: "The fear of the unknown, the dread of the intangible, the unseen terrors that writhe and lurk midst cavernous glooms, all this holds a spice of adventure more poignant than even the claws and tusks of jungle."[12]

Hurley wrote sensational stories for the *Sun*, accentuating words such as "terror" and "dread" to generate the romance of the undersea. But in different contexts, including *Pearls and Savages* (1924), a book published after the film of the same name, the qualities of beauty and pleasure took precedence over sublime terror. Hurley observed how the coral reef is "unexcelled for beauty among all the spectacles of the Universe."[13] Beauty and terror: it was the same opposition that made the underwater such a filmic and photogenic subject for J. E. Williamson. But coral reefs and the underwater also promised an exciting, uncanny vision because the sights that lay beneath the surface were strangely similar to the land's terrestrial shrubs, flowers, birds, and rocky terrains, yet also very foreign. The challenge for Hurley, though, lay in getting beneath the surface.

In 1920, Hurley was not only a national hero but also a man whose reputation for daring and ingenuity had gained international recognition following three major assignments. The first was to the Antarctic with Douglas Mawson on the Australasian Antarctic Expedition (1911–1913), the second with Ernest Shackleton on the Imperial Trans-Antarctic Expedition (1914–1917), and the third as official photographer for Australia in the First World War (1917–1918). Moreover, in 1919, Hurley had filmed Australia by air on the historical Ross Smith flight from London. That he had already mastered an aerial perspective of the world was demonstrated to the public in a film titled *Ross Smith's Flight from London to Australia* that premiered in 1920. The film expanded every citizen's scale of vision from the land to the skies. And with the Great Barrier Reef trip, Hurley intended to expand that vision even further by showing the public the depths of the ocean. His fantasy and dream was to be the one who helped others imagine the scale of the planet.

By 1920, therefore, Hurley could truly say he had managed to escape the jaws of death more than once, to take new viewpoints on the social and physical world, and to innovate new ways of seeing. At the Great Barrier Reef, the Torres Strait, and Papua, he intended to extend an already groundbreaking repertoire of images by seeking out the new and radical underwater viewpoint. With the Ross Smith flight over, Hurley had achieved a bird's-eye view of the earth, and now he wanted to succeed with the fish-eye view as well.

In 1919, Hurley toured Australia delivering a series of "lecture films" about the Ernest Shackleton Expedition in Antarctica, the fate of the ship *Endurance*, and the film he had just completed, *In the Grip of Polar Pack-Ice* (1917). Lecture films were forms of public entertainment in which the speaker, usually a celebrity, aimed to spellbind audiences with entertaining talks supplemented with visual aids such as magic lantern slides and short films.[14] To packed picture theatres Hurley told stories about resilience and fortitude, such as when the *Endurance* was crushed and sunk by polar ice in the Weddell Sea, and when he and twenty-two others were stranded for months on Elephant Island, surviving under upturned lifeboats as they waited for Shackleton, who had gone in search of rescue, to return. It was on Elephant Island that a plan to explore the tropics of the Pacific was hatched: "In the daytime, we talked of nothing but the tropics and the palm trees and while the wind blew ninety miles an hour and the snow covered our shelter, our party planned an expedition into New Guinea the moment we were rescued."[15]

Still on lecture circuit in January 1920, Hurley noted in a diary entry how he suddenly felt "the lure of the sea."[16] By late 1920, he had embarked on the first journey that took in the length of the Great Barrier Reef. That he had left his wife and children at home in Sydney to go on expeditions that lasted months and totaled years is remarkable but also predictable, according to theories of the heroic life. Mike Featherstone argues that heroes typically make such choices. He explains how the myth of the hero is made through a differentiation of the heroic life from everyday life. The heroic life entails the idea of leaving behind the everyday world of domestic affairs to enter one of unfamiliarity and danger.[17] It entails a masculine ideal of putting distance between heroics and the femininity of the everyday. Even in 1915, Hurley was praised as the man who made the unknown known. He was described as "Australia's most daring photographer. He must be ready to leave at a moment's notice for a long hard battle for pictures."[18] As an explorer of frontiers and a self-made man, Hurley constructed his identity by putting the feminine at a distance, and by conceiving of wilderness as a stage on which to act out masculine ideals of "battling" and then return home with evidence in the form of photographs and films. In Hurley's day this was seen as a worthwhile endeavor, but times changed, and by the midtwentieth century the anthropologist Claude Lévi-Strauss wrote in his memoir, *Triste Tropiques* (1955), how he found the enterprise of the explorer-photographer superficial in its aim to capture exotic visual spectacles for popular audiences.[19]

On the journey in 1920, Hurley took an immense load of photographic equipment, including cameras, tripods and glass plate negatives for the production of motion pictures, lantern slides, and newspaper illustrations. The extent and nature of his enterprise typified the modern photographer-explorer. Hurley was meticulous about diary records, and his first entry on December 1, 1920, explained the private purpose of the expedition: to gather material in the form of films and negatives for a travelogue film and lecture-film about the tropics.[20] He planned to screen and publish the future film (later titled *Pearls and Savages*) along with still photographs in Australia, Britain, and the United States, and he envisioned the day when international celebrity speakers would deliver lecture films on his behalf, once the work was released overseas.

The myth started to build that the underwater was a virgin territory, that Hurley would be the first to explore it, and that underwater photography was new. If he could get under the sea with a camera, his prize would be to capture the "unseen" and the "unknown," as colonials were apt to call regions of the planet that were new to them but not to others, such as Indigenous maritime peoples and marine animals and plants. As we have seen with J. E. Williamson, to imagine the undersea as a tabula rasa was a Eurocentric blind spot. It emanated mostly from the heroic tradition of exploration that "holds that a place simply is not a place until it has been visited by a human being"—and more so, until it is visited by a European.[21]

Every time Frank Hurley mentioned in the popular press that he would conquer the underwater at the coral reefs of the Australian Great Barrier, he unwittingly exposed the depth of European ignorance surrounding traditional ownership. It would take the Murray Island Land Case, which overturned the concept of *terra nullius* (nobody's land) in Australia, for non-Indigenous Australians to recognize the Indigenous system of ownership of the reefs and the underwater. Through this case, Australians learned, often for the first time, that the underwater is not without history or culture.[22] When the case, known as the Mabo Case, after elder and plaintiff, Eddie Mabo (1936–1992), was settled in the Australian High Court in 1992, a main issue was the "collective right to the home reef."[23] Ownership involves the floor of the sea itself, including the reef and the less tangible, ever-shifting column of water above the reef.

The idea of the undersea as a virgin territory for the expansion of vision and knowledge helped Hurley to self-mythologize. His ambition in 1920, to dive and photograph underwater at the Great Barrier Reef with a waterproof camera, was timely and relevant to new directions in modern visual culture. However, technologies for underwater photography were still in the process of development. About the history of this specialized area of photography, A. Krista Sykes remarks that it was not until 1927 that taking pictures and films with the body immersed in the sea was a viable practice.[24] As previously noted, a few others had achieved some success, including Louis Boutan.[25]

If, in 1920, Hurley had succeeded with his aim to submerge a camera and film and photograph the tropical underwater in situ, he would have led the field, internationally. The earliest successful underwater filming of coral reefs at the Tropic of Capricorn with a submerged camera is dated around 1928, during the Crane Pacific Expedition (1928–1929) to the region then called German New Guinea.[26] With diving gear and a waterproof camera operated by Sidney N. Shurcliff, the party was successful in securing underwater footage.[27] The Crane Pacific Expedition, though, was inspired by Frank Hurley's journeys to the Papuan region to photograph people, landscapes, coral reefs and the underwater in 1920–22. In other words, the scope and purpose of Hurley's undertaking had made an impression on later scientists. Further, the Crane Pacific Expedition was sponsored by the Field Museum in Chicago. The museum was seeking to extend the Pacific collection for its Marine Hall. While Shurcliff, on behalf of the Field Museum, went underwater in New Guinea in 1929, the Field Museum in the same year employed J. E. Williamson and the photosphere to collect fish and corals underwater in the Bahamas. The Crane Expedition therefore had indirect links to both Frank Hurley and J. E. Williamson. This historical detail puts in perspective the integral place that both Hurley and Williamson occupied in the expanding science of coral reefs practiced by the United States, Australia, the Bahamas, and Britain, and shows how their work and achievements were interconnected, even if neither man was aware this was the case.

At the start of his first journey to photograph the tropical coral reefs of Australia, as he looked across the surface of a calm sea, Hurley described the vision as "a vast liquid mirror."[28] But at daybreak on December 16, 1920, he had his first real introduction to a coral reef. He rowed to a place where reefs lay half submerged and changed his viewpoint from gazing across the water to looking down. He noticed the liminal zone of the sea's surface that separates air from the water below and that shows simultaneously what is above and what is below. The sight of corals breaking through the surface was symbolic and suggestive, and he photographed the same perspective more than once (see fig. 7.1). Memories of the Antarctic returned. Hurley noted how it was paradoxical that in the tropical regions of the Torres Strait, "the ocean was littered with outcrops from the submerged reef & looked much like an ice strewn sea."[29] The sight of the reef lurking beneath the surface of the water released from Hurley's unconscious the memory of submerged ice in Antarctica and heightened his awareness of the hidden dangers of coral reefs. Instantly he was connected with a long European history of doubt and uncertainty about the nature of ocean reef environments, expressed most famously in the journals of captain James Cook.

As Hurley waded through the shallows, looking at the vision below, he said that he "fell entranced; gazing into the magic splendours of a wild morphic dream. . . . It is beyond any pen to describe and having seen beyond one's mental power to comprehend."[30] To try to capture the clarity of water and the vision of submerged corals, he used the "Paget color process," in which a color screen plate and a black-and-white plate are placed on top of each other (see plate 7). Submerged beneath the surface was an unfamiliar world, one he likened to the different logic of dreams and hallucinations. The challenge lay in how to represent such an awe-inspiring sight. He had to concede that no flower garden on land could compare to the dreamworld created by madrepore corals and anemones underwater. It stimulated in Hurley a "surrealist sensibility," even though Hurley did not call himself an "artist" but rather a photographer, and despite the fact that the expedition he was on preceded the movement of surrealism by three years. Dalibor Vesely argues that surrealism, rather than just denoting a movement, represents "a sub-stratum of the whole modern culture."[31] Susan Sontag, in On Photography (1977), also proposed that a surrealist sensibility was pervasive in modern photography, irrespective of the movement.[32] What the sight of tropical reefs, lurking in underwater, sharpened for Hurley was an eye for the surprising and the marvelous, a fascination with the uncanny mimicries of nature, and an attraction to aspects of the nonhuman world that engage space and form in psychological ways. Hurley's diaries suggest that in his encounter with the tropics, and with tropical coral reefs viewed through clear water, he felt liberated from ordinary realities, and it is for this reason that it can be said he discovered a surrealist sensibility.

In 1921, Hurley wrote in a diary entry how "the barrier & its accompanying reefs is one vast playground where nature amuses herself creating fantastic life forms—vegetable & animal."[33] Fifteen years later, André Breton, in *Mad Love*, would define the surrealist aesthetic of "the marvelous" by similarly invoking the uncanny ability of corals to mimic animal, vegetable, and mineral forms. "Experiencing the ocean is similar to life lived in the realm of dreams," observes Natascha Adamowsky, and on the first expedition to the Great Barrier Reef, the Torres Strait, and Papua, Frank Hurley, who had spent much of his life on the sea, discovered a dreamy reality.[34] The sights were like fairyland, and the sea as "calm as a vast sheet of liquid glass & we floating on it as if passing through the glass roof of some fantastic garden."[35]

Hurley never overcame his wonder at the glasslike clarity of tropical seas. But did he try to dive underwater? Nowhere do the diaries suggest that he experimented with his own body submerged in the sea. There is nothing that explains how his skin felt in tropical water. The diaries do not even confirm whether Hurley could dive. Moreover, there are no descriptions of swimming—only descriptions of wading in the shallows. It raises the question of how mobile and capable Hurley actually was in the sea. It is strange for a man who spent so much time on the ocean that there is so little to read in the diaries about the feel of the sea. Instead, there are elaborate and poetic accounts of the sea's surface and of sights that were observable from on deck.

Hurley did, though, watch intently as two Torres Strait Islanders, who were assistants, dived underwater to collect giant clams and marine life that would later serve as camera subjects. As with J. E. Williamson's "native" laborers, the men who assisted Hurley made it easy for him to undertake photography and filmmaking without he himself having to prepare, collect, carry, and operate equipment and machines on his own. As porters, they transported the immensity of photographic equipment that Hurley had in tow. The Torres Strait Islanders who assisted Hurley symbolized the racial underside of European exploration: the invisible workers who enabled travel and adventure to take place but remained an anonymous part of history. The situation was similar to the more general social condition described by James Clifford in relation to the invisibility of "native labor": "Europeans moved through unfamiliar places, their relative comfort and safety were ensured by a well-developed infrastructure of guides, assistants, suppliers, translators, and carriers. Does the labor of these people count as 'travel'?"[36]

When Hurley's Islander assistants slipped over the side of the boat to gather clams for a future photography session, Hurley watched them on the seafloor below from above the water's surface. He had a viewing device with a glass bottom that magnified the underwater, and with this he could observe the men collecting marine life. The Torres Strait divers were legendary for their underwater endurance, being able to free dive for long periods of time.[37] The sight before Hurley unfolded like a movie on a screen, or, as Hurley himself put it, like the spectacle of an aquarium: "I accompanied two of the best men and being armed with a viewing box, a watertight box fitted with a glass bottom, was able to watch operations equally as well below water as above. This magic pane is pushed a few inches below the surface and one can peer down through the water, much like looking into a glorified aquarium."[38]

A glorified aquarium came into view as Hurley looked through a glass aperture and imagined the underwater as a contained space, separate and beyond him. The undersea conceived this way was a giant container of marine life, and he was on the outside looking in. It is unsurprising that Hurley conceptualized the underwater as an aquarium. By 1921, Western perceptions of the underwater were shaped by aquariums, which were understood as optical devices for visual entertainment that brought pleasure and wonder by miniaturizing the sea on land. Hurley's life experience by 1921 was certain to have included zoos and public aquariums, and observing fish and corals in domestic tanks. The films by J. E. Williamson that he had seen were based in aquarium thinking, being shot through air and through glass. Moreover, throughout the 1920s and into the 1930s, Australia was gripped by an aquarium craze. People were mesmerized by tropical underwater scenes contained behind glass. Aquariums were recognized as modern technologies. For example, the American architect Walter Burley Griffin (1876–1937), who lived in Australia in the early twentieth century, designed a house in Sydney in which the ceiling of the dining room was inset with tropical fish tanks to give viewers a novel, modern, and pseudounderwater perspective of fish from below.[39] And by 1922, a number of popular general-release films on the natural history of the undersea were filmed in aquariums. *Denizens of the Deep* (1905) utilized an aquarium to show crustacea, such as lobsters, supposedly in the wild underwater. *Secrets of Nature* (1922) shows fish and marine life ostensibly on the sea floor but in reality filmed in a biology aquarium.[40]

For all these reasons, Hurley was acculturated with aquarium thinking, yet it was something he knew he had to break away from if others were to believe that he had experienced the underwater in an authentic way, from inside it. To call the sea a "glorified aquarium" is to imagine it as an immense space in which animals and objects are kept and exhibited for humans to look at rather than be among. The sea as an aquarium is an attempt to contain the chaotic, unruly, unknowable extent of the underwater within a familiar human spatial metaphor. But the idea of the sea as an aquarium would, in the end, prove too influential for Hurley's approach to photographing the underwater. It would shape the way he envisioned it, encourage him to domesticate it, and ultimately lead him in 1922, during the second expedition to photograph the tropical coral reefs to the north of Australia, to build an aquarium at Port Moresby, Papua, for photographing underwater scenes.

The collection of photographs produced on the second expedition, which show fish and corals underwater but were in fact photographed in an aquarium, are to this day labeled "underwater photographs," a categorization that implies they were taken by a diver submerged in the sea. "Underwater photographs" is also the label Hurley gave them when he deposited the collection in the Australian Museum in Sydney. While they are among the earliest submarine photographs taken of corals and fish at the Great Barrier Reef, the Torres Strait, and Papua, all were photographed in a tank.

"Find a way, or make one!" was Hurley's motto. The motto was published in *Popular Science Monthly* in 1924.[41] The same motto is attributed to the Carthaginian general Hannibal (247–182 BCE) and to the polar explorer Robert E. Peary (1856–1920).[42] When he couldn't find a way to photograph the underwater with a camera successfully submerged in the sea, Hurley made a way, by having an aquarium built. But that was during the second expedition, in 1922. Whereas for the first, in 1920, there was no question that Hurley intended to photograph in situ underwater. The newspapers compared him to Captain Nemo: "Like Captain Nemo, Captain Hurley will discover to us the wonders of the deep, but he will have for his aid what the fictional Frenchman never had—a modern cinema-camera, encased in waterproof steel sheeting and electrically operated by buttons working through water-tight glands. It is proposed to study by this means the undersea life of the Great Barrier Reef."[43]

Hurley's adventures recalled boyhood stories, especially *Twenty Thousand Leagues under the Sea*. Jules Verne had so shaped Hurley's image of the Torres Strait that in 1924, when thinking back on his time there, he said he had felt as if he was "living in some uncanny drama created by the imagination of Jules Verne."[44] In the Torres Strait, he encountered all the elements of Verne's adventures: savages, pearl divers, pearls, sharks, and the psychology of the underwater. He felt as if he were reliving the novel in person. He had a strange sense that life was imitating art.

Recently Helen M. Rozwadowski has argued that novels, including adventure stories, have shaped many preconceptions of ocean environments.[45] And Sean Brawley and Chris Dixon also observe, "The principal means by which notions of the South Seas were transmitted from generation to generation was through literary texts."[46] It has already been said that Jules Verne was important for shaping ideas of the Torres Strait region. But there were other likely sources. It is striking, for example, to think how similar Hurley's expedition to the Great Barrier Reef, the Torres Strait, and Papua was to a fictional expedition in *Guinea Gold, or, The Great Barrier Reef* (1883), a novel by Charles H. Eden. The protagonists follow the same sea route as Hurley: along the Great Barrier Reef, past Murray Island (Mer) and Darnley Island, to the Fly River in Papua, where they find Papuans who were ill-treated by Europeans.[47] So, while in 1920, Hurley said he was setting off to "unknown Papua," Eden's book demonstrates that the paths through the region were quite well worn by travelers, explorers, entrepreneurs, anthropologists, writers, and artists, including Eden.[48]

Throughout 1921, Hurley experimented with ways to film the underwater. But the experiments were from above water, and in the air. He tried using a casing to hold the camera over the side of a boat with a "gap for cine" to take reef pictures.[49] The device was clipped to the boat. It had a recession, or gap, in which to place the camera with the lens pointing down. The idea was to film through clear water without getting the camera wet.[50] The device would take Hurley one step closer to the undersea than older methods deployed by photographers such as William Saville-Kent, who photographed the Great Barrier Reef in 1893 with a camera on a tripod pointing down through shallow water (see fig. I.1).

In the end, though, Hurley was forced to use an aquarium. Does this matter? It does, and one reason concerns the different optical, physical, and psychological experiences generated by photographing the underwater from air through the glass of an aquarium, compared to photographing with a camera submerged in the sea. It involves the difference between absorption and distantiation.[51] Absorption involves participation. When a photographer is under the water filming marine subjects, it reduces the gap between subject and object. But a photographer in air, look-ing through the lens of a camera as well as the refracting barrier of a glass aquarium to an underwater scene beyond, is disembodied and distanced from the world under study. The photographer is not part of the same me-dium or atmosphere. When a photographer uses an aquarium to simulate an underwater scene, what he or she does is create a kind of diorama.

Hurley always made anything posed, constructed, or artificial look nat-uralistic. When the tropical underwater posed a significant obstacle to photography, he resolved to simulate the effect. As Hillel Schwartz points out, film, in particular, "was a medium for duplicities."[52] Paradoxically, while the new and the original were hallmarks of modernity, along with the persona of the hero and the concept of a frontier, an equally dominant impulse in modernity was toward "the culture of the copy," one that cel-ebrated the reproduction capabilities of the machine age. Both cultural spaces—a culture of the original, the frontier, and the hero, and a culture of the copy—informed Frank Hurley' imagination.

Hurley undertook his second expedition to the coral reefs of Torres Strait and Papua in 1922 in the company of an ichthyologist from the Aus-tralian Museum in Sydney, Allan McCulloch (1885–1925). On returning to the region with McCulloch in 1922, the visual and psychological impres-sion for Hurley was no less hallucinogenic or dreamlike. A diary entry notes, "Nothing in the world has so enraptured me as the sublime won-ders of the coral reef."[53] On the earlier trip in 1921, Hurley had collected mental images of how an ideal photograph or film of an underwater coral reef, with fishes, might look. In 1922, he intended to make those ideas con-crete. At the end of the trip, the *Sun* newspaper published a photograph of tropical coral reef fish swimming underwater. The headline reads, "What the Camera Saw on the Ocean Bed."[54] There was one grain of truth: the floor of an aquarium can be made to resemble the ocean bed. But the in-tervention of the aquarium was something Hurley ensured remained in-visible to the public.

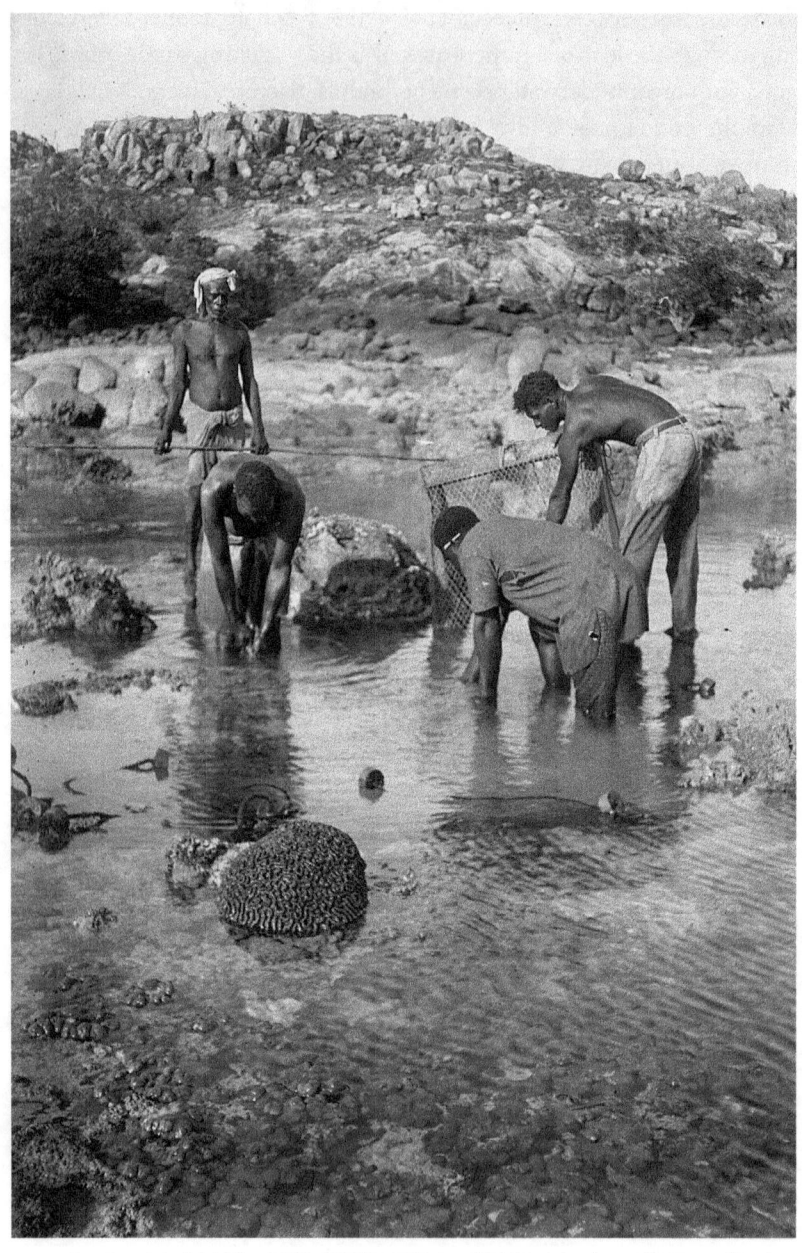

Fig. 8.1. Frank Hurley, "Men Collecting Coral Fishes on the Home Reef,"
1922. Courtesy of Australian Museum Archives, Frank Hurley
Photographs AMS320/V4696.

CHAPTER 8

Hurley and the Australian Museum Expedition

The image portrayed of Frank Hurley by the media, and by Frank Hurley himself, was a mass of contradictions. At times, he was a picture-sorcerer rendering poetic scenes of wonderlands among corals and reef fish. Sometimes he believed that Creation was more reasonable than Evolution as an explanation for the beauty of the reefs. At other times, he was a scientific rationalist who upheld Darwin's theories, praised the bureaucratization of so-called primitive parts of the planet through colonization, and was in awe of the metropolis. He said he adored the wild child of nature, but also wrote excitedly about engineering and the capabilities of the machine in the industrial, capitalist world. He was both a creator of the magical aura of cinema, and a producer of the endlessly reproducible objects of mass culture. He worked in a space in which reason and imagination deliberately played off each other, especially science and myth, knowledge and fantasy, a realm described by Michael T. Saler as the ironic condition of modernity.[1] For anyone who understood modernity through the negative effects of rationality and industrialization, disenchantment was a common trope in the early twentieth century. But Frank Hurley reconciled enchantment with a positive attitude toward technology to construct imagery intended to delight and astonish general audiences with the wonders of nature.

After the first expedition along the Great Barrier Reef in 1921, an expedition that resulted in the release of the travelogue *Pearls and Savages*, Frank Hurley felt there was rich potential for new footage and photographs "to further augment" the first iteration.[2] In 1922, in collaboration with the Australian Museum in Sydney—the first natural history museum established in Australia—he and scientist Allan McCulloch planned a three-month venture along the Great Barrier Reef, through the Torres Strait to Papua, with the ultimate goal of exploring the inland of Papua, including the Fly River and Lake Murray. They wanted to observe, collect, and record people and customs, and undertake other anthropological investigations. The moving images and still photographs they took are analyzed in chapter 9, which also explains the coral reef scenes in the second iteration of *Pearls and Savages*. This chapter, though, interrogates the significance of taking an aquarium to Papua during the Australian Museum expedition of 1922.

As the second expedition progressed, Hurley made records in the form of films, photographic plates, sound recordings, and diary notes, and for the public he wrote newspaper articles. His photographs and films of the lives of island peoples and their customs, rituals, and traditions were punctuated by contrived and dramatic scenes exaggerating the meeting of the modern world, a world that he represented, with the so-called primitive world he had gone in search of. At every opportunity, he created theatrical contrasts of the modern and the primitive. The most famous of these is a juxtaposition between a seaplane (with which Hurley intended to photograph the region by air) and a dugout canoe. A diary entry provides insight into the significance of the symbolism to Hurley: "The amazing spectacle of the native canoes & man's supreme achievement, the aeroplane."[3]

The journey in 1922 was both a scientific and a picture-making expedition. It was to Hurley's advantage that science and picture making were seen to serve each other. Science gave his work in film and photography greater credibility through its association with truth and knowledge. It was especially important to Hurley to secure an official affiliation with the Australian Museum and to have Allan McCulloch along to advise on natural history and anthropology. Robert Dixon points out that until 1922, Hurley "had as yet developed no formal connection with a scientific institution such as a natural history museum."[4]

However, while photography was recognized as a scientific tool, film was treated suspiciously by scientists. Members of the science community looked down upon film, seeing it as a form of mass entertainment. Film epitomized what many saw as the superficiality of spectacle, and questions were raised about its value to scientific work. But as Lynn White Jr. explains, what was at risk was a matter of class. Underpinning the efforts scientists made to differentiate the purity of science from an emerging tendency in modern science to incorporate technologies associated with mass entertainment was the sense that film was "low brow." White explains how "science was traditionally aristocratic, speculative, intellectual in intent," and "technology was lower-class, empirical, action-oriented."[5] So, while, from Hurley's point of view, the collaboration with the Australian Museum was a reconciliation of rationality and enchantment, from the science community's perspective, the films he produced as scientific investigations of "primitive" peoples, particularly *Pearls and Savages*, were dubious. Later, in America, Hurley would encounter some opposition when *Pearls and Savages* was screened, a point discussed in chapter 11.

Science aside, Hurley's objective with the second expedition was also to make money and to "exploit the films throughout the world" so he could pursue further adventures independent of financers and sponsors.[6] On the first expedition, he had sensed a profitable future working with the stunning backdrop of a coral reef environment. Although the region he traveled through, especially the jungle areas of Papua and New Guinea, was known as one of the planet's darkest and most primordial regions, among coastal coral reefs Hurley recognized an opportunity sympathetic to photography—a flow of air and light that typified modernity's fascination with clarity and transparency.[7]

Along rivers that Hurley imagined were "hitherto untraveled by whiteman," and that he fantasized as harboring a precolonial pure form of culture protected by wilderness that was still intact, he and McCulloch collected insects and shot and skinned birds-of-paradise for the Australian Museum.[8] Did they see the irony in colonizing and destroying the very wilderness they romanticized? With them they carried wireless apparatus to keep in touch with European settlements and send reports back to the Australian mainland. It was an ill-fated trip: in 1923, there was a suggestion that Hurley and McCulloch had stolen sacred objects and placed "immense difficulties in the path of the [colonial] Administration, which was endeavoring to bring the wild natives under control."[9]

It was in 1922 in Port Moresby that Hurley had the aquarium built. Larger than the average home aquarium (50 × 30 × 30 inches, or 127 × 76.2

× 76.2 centimeters) it came with a removable, sloping back "for displaying the exhibits to advantage."[10] The concept of having a sloping back with a vertical accent was based in the theory of optical illusions. The angle of the back was a perspectival device for intensifying the effect of abundant marine life and deep space in what was otherwise a confined and shallow area crammed with marine animals. Palle Petterson points out that this was a time when photographers and filmmakers "had an incredible talent for composition and, when necessary, they 'staged' nature in order to capture the scenes they wanted. They made sure, however, that the result appeared 'natural.'"[11] Although Hurley's attempts at diving and taking photographs underwater had failed, there were real advantages to using an aquarium, not least of which was the way it enabled observers to get very close, and at eye level, with marine life.

Hurley said the effect he was after from the aquarium display was a close resemblance to a coral pool, the type found in shallow waters that travelers and tourists like to gaze into, contemplating the meaning of life and the picturesqueness of nature (see fig. 1.1). He wanted to reproduce the same allure as realistically as possible.[12] It was a compromise to use an aquarium and create an artificial underwater, but there was much satisfaction in knowing that the fish and corals he photographed would be authentically of the tropics, captured and filmed as close as possible to their own environment. As such, the results would be based in truthfulness and accuracy, and the aquarium would lend the additional dimension of enchantment.

But, however he rationalized it, it was unusual to take an aquarium to the home of corals and tropical fish with the intention of removing them from the sea to photograph them on the shores of a coral island. It was especially unusual since Frank Hurley had pledged to the public only a year before that he would pioneer the filming of the "floor of the sea," that he had diving gear and underwater cameras and that with this equipment he would film and photograph tropical coral reefs and fish in situ. Every report insinuated that he would get under the surface and be part of the medium of seawater. What the aquarium guaranteed was that Hurley would not go home empty-handed.

In Port Moresby, Hurley hired a canoe and a Papuan crew to transport a team of explorers and scientists, the aquarium, and a huge store of camera gear on a camping tour of local islands. In advance of setting out for the remote Fly River, he intended to film and photograph coral reefs and fish on the islands close to Port Moresby. For this he needed Indigenous guides. In this region of the southwest Pacific Ocean, where Australian as well as British, French, and German colonies were established, the labor of island peoples as navigators and porters was essential for westerners working there.[13] In 1921, Hurley admitted that "life here [in Papua] would be impossible without the natives to do our carrying, hardwork & menial duties."[14] Yet white and Indigenous groups were routinely segregated. An example of racial segregation in the Pacific region in 1913 is given by Arthur Grimble. Writing in 1952 about work he did much earlier for the Colonial Office in the British protectorate of the Gilbert and Ellice Islands (Kiribati and Tuvalu), to the northeast of Papua New Guinea, Grimble remembered the strict divisions between peoples in islands that were known as "colonial possessions," where colonialists acted like prefects, and Indigenous peoples were "the school-children of Empire."[15]

The circumstances surrounding the transportation of Hurley's aquarium makes for a striking and symbolic vision: a Melanesian canoe and crew in the colonial tropics transporting a Western fish tank over reefs teaming with fish and brimming with corals. The vision is every bit as surreal as the scene in Jane Campion's film The Piano (1993), in which a piano, having been transported from Scotland in the mid nineteenth century, sits stranded on a beach on the west coast of New Zealand, conspicuous and imperial. On a canoe, the aquarium sat, symbol of Western civilization, the Old World, and the colonial practice of actual and symbolic acquisition of territory. In it, the creatures of the coral reef would become captive in their own land, rather like the Papuan crew, subjugated by colonial administrators.

On reaching the small island of Dauko, Hurley sent the crew of men who were employed to transport, navigate, and provide labor to the reef to collect fish from rock pools and noted, "There is little that the natives do not know about fish netting & the reefs" (see fig. 8.1 and plate 8).[16] McCulloch, though, was more adventurous, and took the opportunity to borrow a pair of tortoise-shell goggles from the crew. He dropped over the side of the boat and watched as the men below set nets to catch fish. He remarked on how the glasses had originally been introduced by Japanese pearl shellers "to enable them to see clearly under water."[17] McCulloch's recording of ethnographic information about Islander interactions with the sea was part of an increasing interest in learning more about environmental practices in coastal maritime communities.

In addition to ethnographic enlightenment, McCulloch and Hurley were also fascinated by the spectacle of Indigenous peoples in the underwater—exotic space for exotic bodies. European eyes racialized coastal peoples in tropical seas and at coral reefs by characterizing the relationship as natural rather than cultural. Nicole Starosielski also makes this argument in relation to the Bahamas and J. E. Williamson's cinematic vision of the 1910s that "saw the subaquatic as the domain of an ethnic Other. It dramatized the labor of 'native' bodies under the sea, but contained their power through comparisons with aquatic animals."[18]

McCulloch didn't dive underwater; instead he looked through goggles from the surface at the divers in the depths below, taking in the whole scene with his eyes as if filming it. From that distance, and behind the glass of his goggles, it is easy to imagine McCulloch's gaze objectifying the strangers below as if he were seeing them in an aquarium. Yet by observing the skills of the divers as they engaged with the language of the sea, McCulloch surely also realized that the underwater was not a virgin territory for representation, as Hurley had claimed through the media, nor a space empty of history and culture or a submarine *terra nullius* belonging to no one, as westerners were apt to presume. Rather, it was owned by others, already modified by technologies such as goggles and fishing nets, and, for Papuans, part of the culture of everyday life.

After sending the Papuan crew to the reef to collect fish and corals for the aquarium, Hurley took photographs of the men at work. In figure 8.1, they can be seen combing rock pools for small fish, an incongruous sight for men who usually sailed the reefs to fish for food rather than collect fish for display. Figure 8.1 is not a romanticized image, unlike plate 8, which is studied and contrived and perplexing for the way the fish are laid out on a mat in the air without water. By contrast, the men in figure 8.1 are not cheerful, and the scene is very different from Hurley's famous portraits of Papuan "savages" showing nakedness, ritual, and decoration as markers of cultural otherness and the timelessness of "primitive" life.

A photograph of Hurley and McCulloch relaxing on Dauko Island surrounded by the Papuan crew, with the aquarium in the background, was reproduced in the *Sun* newspaper in 1922 (fig. 8.2). Titled "With their dusky friends," there was no article to accompany the photograph, and no explanation of the strange box in the background. Beneath the image, though, a caption described "Captain Hurley and Mr. McCulloch listen-

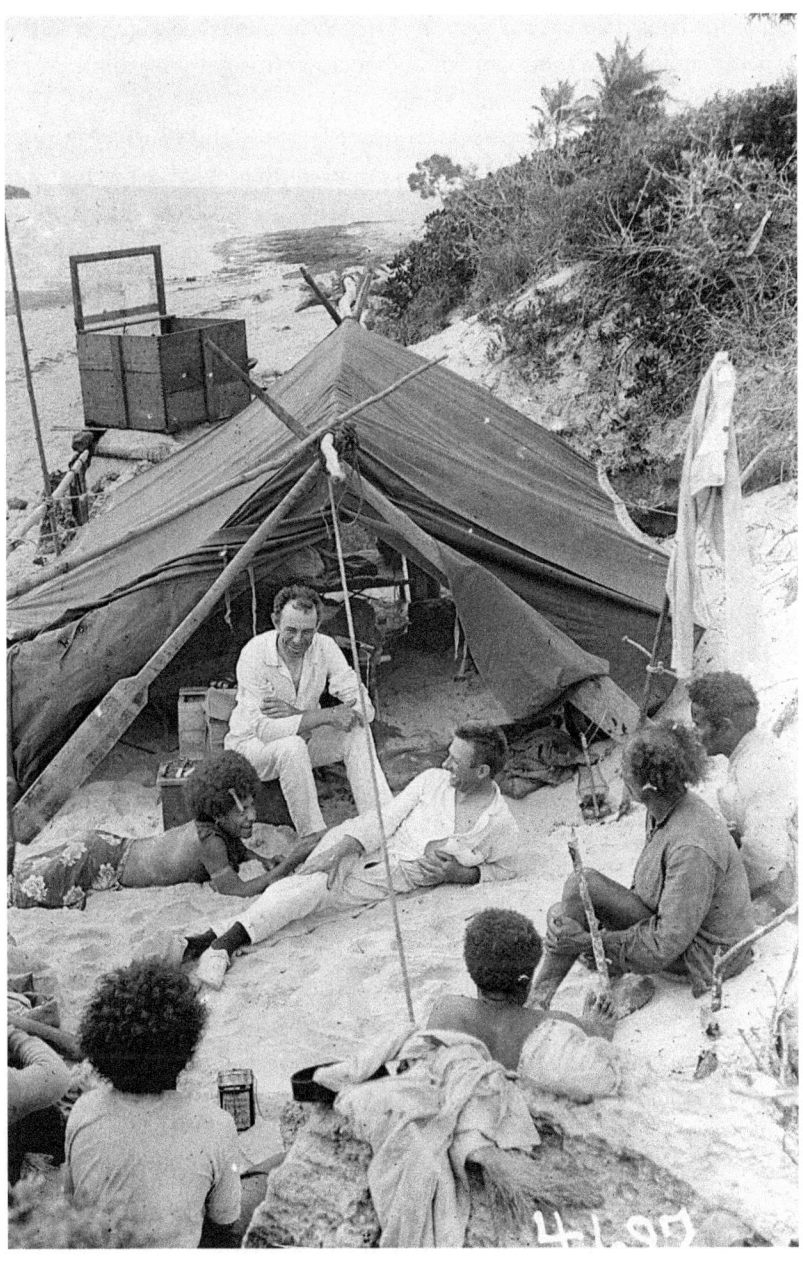

Fig. 8.2. "Hurley, McCulloch Dauko," 1922. Photographer unknown.
Australian Museum. Courtesy of Australian Museum Archives, Frank Hurley
Photographs AMS320/V4697.

ing to the latest Papuan story."[19] The image was published to promote the expedition and give metropolitan audiences an insight into cultural and racial difference in the colonial tropics.

The photograph gives the appearance of meaningful contact between Hurley, McCulloch, and the Papuan crew, yet there is also the familiar contrast often seen in ethnographic photographs of the tropical whites of explorers and scientists next to the black skins of the Papuans. The groups appear to be comprehending each other, sharing a joke, and enjoying the moment, and give the impression they are collaborating on a joint adventure. There is a distinct atmosphere of leisure and play in the image, as if the group has gone on a camping holiday to swim, sail, hunt, and fish rather than undertake a scientific expedition. Helen Rozwadowski observes that early modern scientific fieldwork often resembled "camping, hunting, [and] fishing" trips, and that for "the scientist themselves, fieldwork and play were indistinguishable."[20] The same holiday atmosphere can be observed in scenes captured by J. E. Williamson in *Under the Sea*, the film that records the scientific expedition he conducted with the Field Museum in which the camera records the African-Bahamian crew eating, swimming, and playing on the coral sands of Sandy Cay.

In his diary, Frank Hurley wrote how he lay on the beach with McCulloch and the Papuan crew. He said he felt close to nature "and glad to get back again to primitive life."[21] Cheerful talk of the primitive life was something Hurley envisioned as a good insert in a newspaper column or in a lecture. But despite the photograph's sense of racial integration, Hurley and the white travelers usually separated themselves from the Papuan crew. Even the food rations reflected social position—a separate list of items, headed "native rations," itemized basic foods, such as rice, herrings, solid meat, and sugar, while another, headed "whites," reflected the exotic tastes of empire: lemon butter, curry powder, Bovril, coffee, and New Zealand sheep tongues.[22]

While Hurley respected the Papuan crew for their local knowledge, he also saw them as primitives and children who were related to him—but very distantly. Setting up the aquarium on Dauko Island, he noted in his diary how he and McCulloch "entertained the natives with talks of cities & civilization & they in turn told us many things of their own simple life."[23] He had somehow connected Jean-Jacques Rousseau's idea of the savage and child as innocents of nature with social Darwinist theories that primitives and children are evidence of human development at its most rudimentary. He felt superior and paternal but also longed for a return to primitive simplicity: "If I were a savage—without ambition—which after all counts for most of the miseries of life—I would not seek to be more than what I was—a wild free child of nature rather than a serf of civilization."[24]

But being a man of many contradictions, and having a range of different audiences to address across the realms of popular culture, art, and science, it was difficult for Hurley to maintain one voice. Within a few months his mind had wandered away from a longing for primitive nature to a longing for civilization. In November 1922, on Dauko Island, he heard the sound of the sea breaking on the reefs of the Pacific and said he recognized in the sound "a time old anthem—a song of Empire."[25] Many Australians had abandoned their Britishness after the First World War, but in the sounds of waves crashing on reefs, Hurley was reminded that Britannia ruled the waves of a vast global empire. The ocean was the force linking all the empire's lands; it was a significant symbol in what Robert Dixon calls the imperial imaginary and the "occupation of a global space."[26] In remote Papua, the sound of waves gave Hurley a sense of connection to every other British citizen.

In its own way, the aquarium was also another symbol of the occupation of global space. Colonized lands could be miniaturized within its walls. For Hurley, though, the aquarium was primarily an optical device for photographing captured tropical animals that he hoped would provide him with profitable imagery. With time running short and their departure for the Fly River imminent, Hurley, McCulloch, and the crew worked hard to find a way to obtain beautiful pictures of idyllic coral reef life from the viewpoint of the underwater. But something unforseen went wrong with the aquarium. Each time they arranged corals and fish in preparation for photography and filming, the corals died in the tank and emitted a slime that interfered with Hurley's effort to obtain crystal clear images. The combination of stagnant tank water, the collection of silt in the aquarium, overexposure to light, the heat of the sun, and the depletion of oxygen by fish and corals packed into such a small space turned the aquarium into a toxic setting. In 1922, Hurley had unwittingly created an environment that produced the same biological degredation that in the twenty-first century is known as coral bleaching, the disaster that now threatens the future existence of the reefs that Hurley traveled across and photographed in 1921 and 1922.

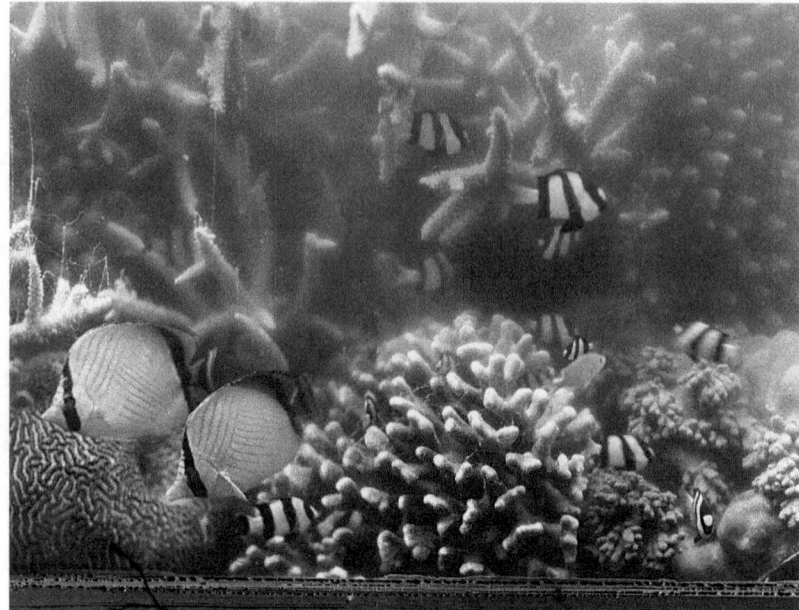

Fig. 8.3. Frank Hurley, underwater scene, Papua, 1922. Courtesy of Australian Museum Archives, Frank Hurley Photographs, AMS320_V05093.

But his immediate problem was the way the slime emitted from dying corals spoiled the picture. The aquarium had to be emptied and refilled again and again with fresh coral stock.[27] One surviving image records the scene of corals emitting slime under stress in the aquarium (fig. 8.3). Threads of slime can be seen rising from the corals on the left. What we see is the expulsion of zooxanthellae, a reaction that leads to "bleaching."

Frank Hurley and Allan McCulloch made notes about the "mucus" and "slime" that the coral stock emitted. As naturalists, they were in the habit of considering and evaluating their actions, including the loss of specimens. If anything, they seemed dispassionate about sacrificing corals and fish in order to progress scientific knowledge and obtain spectacular images. Not knowledgeable about corals, Hurley had to feel his way toward an understanding through trial and error: "The corals, especially the madrepores, were still giving off their slimy mucus, so that the tank had to be emptied & refilled. This necessitated carrying 1,500 lbs of water from the beach in tins. Arduous work in which the natives cooperated, though had it not been for McCulloch I doubt if I could have maintained sufficient enthusiasm to carry through."[28]

In 1922, there was but a rudimentary understanding of human impact on coral reefs. This was ten years before governments sought to protect the Great Barrier Reef. And not for another six years would the British scientist Charles Maurice Yonge, who teamed up with the Australian Museum from 1928 to 1929 to work at the Great Barrier Reef, first theorize the metabolism of corals and the existence of "zooxanthellae, the single celled algae which reside in the coral's gastroderm" and that are driven out in slimy strands when corals are under stress, leading to "bleaching," since it is they that give corals their color.[29]

Although Allan McCulloch was a scientist, his concern with the slime emitted by the dying corals was the same as Hurley's: it interfered with the production of sensational images. So with their hearts set on obtaining clear images of tropical fish swimming underwater, Hurley and McCulloch removed the corals, scraped away the dying coral polyps, put the remaining structures in the sun, and bleached them until they were white skeletons. McCulloch wrote, "We set our aquarium tank on the beach of Lolarua and arranged coral and put the fishes into it to take cinema and still pictures. Its width of 3 feet was found to be too great, so it was divided with a large board, but the corals gave off so much slime that the water quickly fouled and obscured all but the objects near the glass front. . . . In the end fine cinema pictures were obtained using only bleached corals."[30]

Today the image in figure 8.3 is remarkable in a way Hurley did not anticipate—as a portent of things to come for the world's coral reefs. Today, the photograph returns a gaze about the anthropogenic conditions of the present environmental crisis of coral reef ecologies. So while we can look at Hurley as an early advocate for promoting the natural wonders of coral reef environments, it is also the case that the patterns of engagement we read through his diary entries, about collecting corals then throwing them away when they started to die, suggests an attitude that nature was expendable and an endless resource. That the natural world was a limitless resource was the prevailing view of modern times. Today the Great Barrier Reef faces extinction.

Taken through the glass front of the tank, the image in figure 8.3 has a cloudy atmosphere that interfered with the perfection and clarity that Hurley was after. But the image did have its uses. In 1928, it was published in an art book, a small book written by Randolf Bedford on the Great Barrier Reef for the publishing company Art in Australia.[31] In the context of an art publication, the photograph's atmospheric qualities worked to Hurley's advantage. At a time in history when the artistic style of pictorialism was current, the gossamer threads of slime that signify dying corals also evoke a fairyland dream-space of atmospheric effects and hallucinations. Soft, diffuse, suggestive, and mysterious, the image in figure 8.3 was perfect for art.

When published in Bedford's book, the caption below the photograph in figure 8.3 told readers, "Coral growths under the sea. Great Barrier Reef, Queensland. The stillness and clarity of the water enable photographs to be taken of the marine life among the coral structure of the Reef."[32] In the expanding world of visual media of the 1920s, photographic images were often misrepresented with captions that were either intentionally or unintentionally misleading about place and circumstance. Consequently, there was no mention that the photograph in figure 8.3 was taken on the shores of an island in Papua, or that it was photographed through the front of an aquarium. Instead the caption suggests the photographer was submerged underwater among living corals in situ, rather than observing them in a tank. Beford's book was published at a time when Australia made a concerted effort to broadcast the beauty of the Great Barrier Reef and promote tourism in the Queensland section of the reef. To mention Papua, a place associated in the Western mind with "headhunters," had the potential to intimidate future travelers. Therefore it was expedient to switch the geography of Papua for Queensland and to avoid mention of the aquarium. It was certainly better for Hurley's image not to mention the aquarium, because it intensified an already held view that no part of the planet was too difficult or remote for the explorer Frank Hurley.

One image from the photography session that took place on Lolorua Island in 1922 stands out. It is a photograph (plate 9) that Hurley himself found more effective than others. Its visual impact is quite different from the effect he described on the first expedition in 1921. In 1921, he looked over the side of the boat to the submarine world below and witnessed a scene chaotic, kaleidoscopic, and mosaic. A diary entry recorded a sight in which "neither order nor rule controlled the color scheme of this profound camouflage."[33] Yet in 1922, the effect of chaos and camouflage was replaced by an aesthetic decision relating to audience appeal. Instead of camouflage, Hurley sought heightened visibility for tropical fish swimming among corals. The aquarium allowed him to push everything to the front of the glass. Blending and hiding were the opposite behaviors he wanted from camera subjects that would bring in profits. Camouflage was the antithesis of making objects stand out, and therefore unhelpful in the production of vivid images of tropical life. Like many owners of domestic aquariums, Hurley was after the discernibility and conspicuousness of animals going about their captive lives for people to watch.

Frank Hurley turned the photograph in plate 9 into a colored lantern slide for lectures. In 1924 the image was reproduced in the *Illustrated London News* and in the book *Pearls and Savages* by Putnam's Sons.[34] By coloring the image—placing a colored plate as a screen in front of a black-and-white plate before exposure—Hurley was able to grace the final vision with a more seductive beauty than black-and-white would allow.[35] The year 1922 was on the cusp of a new era for color photography, especially for moving pictures. Not until 1924 did filming in color present itself to Allan McCulloch as a practical and realizable method for filming and photographing fish, but he was excited that "soon it will be possible to illustrate lectures with movies showing both the form and colour of the brilliant tints of fishes, and their many other brightly-hued associates of a coral reef."[36]

On Lolorua Island, McCulloch and Hurley shared ideas on the best way to photograph fish underwater and on the special effects they sought. The performance and beauty of the animals was important to Hurley, since a main goal was to impress the public with startling new images.[37] But the behaviors, responses, and reactions of the Papuan crew was also something Hurley liked to study. He watched the men reacting to fish swimming in the aquarium, observing how the fish that were so familiar in a wild state became unfamiliar in the miniaturized ocean of the aquarium and by the magnification of glass: "Particularly interested were the natives who though they had seen the fishes in their natural home, were deeply impressed when they were isolated, & could contemplate them without the distracting influence of the great bewildering expanses of the seabed."[38]

It was as though Hurley were excited by the idea that in that moment of observing captive fish in the aquarium, the Papuan crew had been Westernized and civilized. Exposing them to corals and fish in the aquarium was like taking them to an art gallery to contemplate autonomous works of art. They had become like spectators of pictures and objects of contemplation; they could appreciate beauty in a disinterested way. The aquarium, as a miniaturization of the ocean, enabled their eyes to focus on delight in detail and smallness. Hurley also imagined how the aquarium enabled the Papuan men to contrast an aesthetic experience of beauty contained within the tank with the sublime, wild ocean bed that he described as so vast and abstract it engenders only bewilderment. With an aquarium, the eye can rest, whereas the sublime expanse of ocean is, as Alain Corbin explains, a "spectacle [that] prevents any rest, especially for the eyes."[39] Hurley seemed pleased with himself that he had transformed an Indigenous concept of nature into Western terms: nature as a picture.

The scenes Hurley created in the aquarium were discrete, idyllic, underwater visions of exotic animals. But they were also more than this; symbolically, they were scenes of "imperial possessions" that signified European mastery over tropical nature.[40] Hurley's aquarium was similar to zoos of the Victorian period in which, as Susan G. Davis explains, "animals were interpreted in explicitly imperialist terms; as tokens of conquered peoples and metonymic extensions of the geography of empire, they were illustrations of racial inferiority and difference."[41] Aquariums by their very nature and size involve the act of compressing undersea animals into a "chamber of wonders," and for Bernd Brunner, this is an act of colonization because it establishes control over other beings.[42] In the broader field of travel and tourism studies, John Urry makes a parallel argument about the processes of colonization in relation to exotic cultures, in which he claims that the object of the gaze is tamed through the process of looking. Momentarily at least, to have visual knowledge of an object is to have power over it.[43]

Eventually, still photographs from the photographic expedition to Dauko and Lolorua Islands were placed in the *Illustrated London News*; a popular science book, *The Great Barrier Reef* by zoologist William Dakin (1950); *Popular Mechanics* magazine, in which science was made accessible to the everyday reader; and Randolf Bedford's *The Great Barrier Reef*, the first art book in Australia on the subject of the reef. Some film footage was blended into the amended version of the travelogue *Pearls and Savages* (1923), a film that was screened in the world's museums as well as local picture theaters. Notable for a five-minute interlude in which a camera lingers over a scene of a living coral reef, and arguably the first general-release film to incorporate the natural history of the "wonders" of the Great Barrier Reef, *Pearls and Savages* combines moving pictures taken above water in the Torres Strait with moving and still images of fish and corals shown beneath the surface. The film was the realization of Hurley's dream to belong to a distinguished community of explorers and scientists, and to create and profit from the most sensational images the world had ever seen.

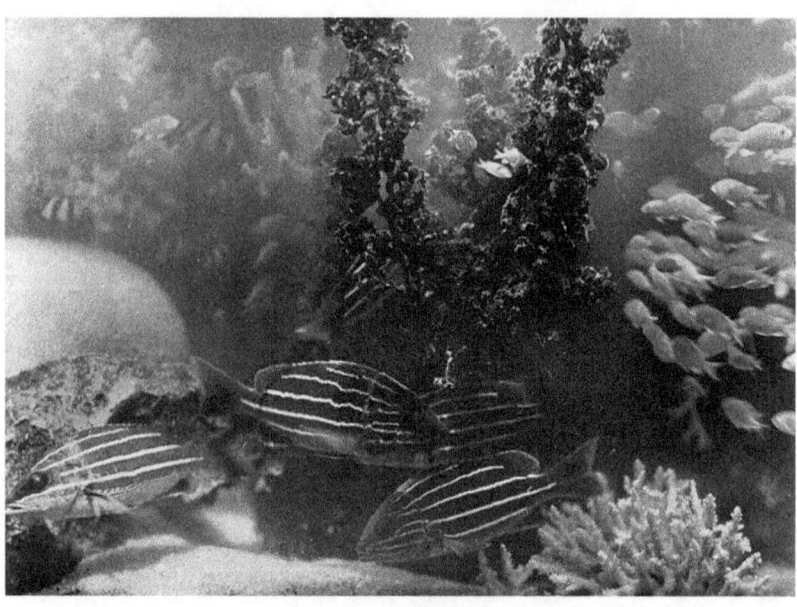

Fig. 9.1. "A submarine photograph by Captain Frank Hurley, showing the dense marine life typical of the Pacific Coral Reef areas." From Randolph Bedford, *Great Barrier Reef*, 1928, unpaginated.

CHAPTER 9

Pearls and Savages

Pearls and Savages was the product of a colonial modernity in which visual culture was underpinned not only by colonialism but also militarism and patriotism.[1] Celebrated as an informative and imaginative representation of empire, the film was circulated widely to audiences seeking entertainment from the exhibition of strange lands and strange peoples. According to Erik Barnouw, early modern documentaries about the remote regions of colonial empires typically reflected "coverage of 'natives' [that] generally showed them to be charming, quaint, sometimes mysterious; generally loyal, [and] grateful for the protection and guidance of Europeans."[2] In that regard, *Pearls and Savages* is typical of the documentaries of colonial modernity, since the subject of "natives" is central to a film that is primarily about Frank Hurley's travels through the Torres Strait into "deepest Papua" to find the legendary "cannibals" and "headhunters" of the jungle. Vital is the element of native mystery and quaintness, along with drama, suspense, and surprise. But sandwiched between ethnographic scenes filmed in the Torres Strait of men fishing and capturing turtles is a five-minute interlude of wildlife photography of coral reefs at low tide. In place of human action we see marine animals in coral pools on the reef and in the underwater. It is footage rarely talked about in film history, in which the human component of the film usually dominates. Camerawork focuses on the details of coral formations and pools, and on animals that, as the intertitles inform us, are strange beings. The camera moves slowly and invites spectators to feel surprise and admiration for one of the world's greatest wonders.

Because postcolonial discourse focuses on human interactions and power relations between colonizers and colonized, the discursive field in which Hurley's *Pearls and Savages* is situated has ignored the natural history component of the film detailing the reefs and the underwater of the Torres Strait. Kathryn Ferguson observes how films of the underwater, and by extension the nonhuman world more broadly, suffer in terms of critical discourse because documentaries in which the human figure is absent, as distinct from dramas involving human subjects, do not generate the same level of interest.[3] Yet human and nonhuman natures inform each other in *Pearls and Savages*. The idea of splicing displays of fantastic-looking sea creatures, such as sea urchins and wild-eyed tropical fish, between scenes of exotic people, was an alternative way for Frank Hurley to register the entire human and nonhuman environment of the Torres Strait and Papua regions as wild and strange to him and to the audiences intended for the film. The entirety of *Pearls and Savages* is a portrait of strangers and strangeness and qualities of otherness. To Western spectators of the 1920s, visions of "natives" seen in the context of native flora and fauna symbolized an alien world beyond their sense of home and familiarity. In Hurley's film, the combination of "natives" and marine wildlife represented an alterity that viewers were invited to observe but also to keep at a distance through the objectifying medium of film and the spectacle of the screen.[4]

The film, however, was just one module of a larger event and performance intended to dazzle audiences. And dazzling the audience through performance was something Hurley had learned from two sources: early twentieth-century Western performers who delivered illustrated lectures to public audiences, and performers in villages in Papua in 1921, including the dancers of Buna, who held him enthralled with their "changing & flashing colors [like] a glorified human kaleidoscope that dazzled the eyes."[5] To increase the spectacle of *Pearls and Savages*, Hurley's illustrated lectures involved evocative film screenings, vivid stories, and the kaleidoscopic effects of light projections through magic lantern slides as they dissolved into one another. And in addition to the production of the film, there were collections of still photographs, promotional literature and advertising, magazine articles and newspaper stories, books, and catalogues.

The film itself went through a number of name changes. It was released in Australia in 1921–22 as *Pearls and Savages: A Film Drama of Primitive Humanity*, and again in expanded form in 1923 as *Headhunters of Unknown Papua*. The iteration of 1923 was based on material collected during the expedition with Allan McCulloch in which additional footage was gathered, including images of fish filmed in the aquarium that Hurley had built at Port Moresby. In 1921, Hurley felt encouraged by the positive public reception of the first iteration of the film, and decided to set out again to acquire material he had originally discounted or had found difficult to obtain. A newspaper review of the release of *Headhunters of Unknown Papua* in 1923 announced that it had been Hurley's intention to "open all the sealed pages in the book of Papua," implying again that the region was unknown to outsiders, when in fact it occupied a fairly well-worn path, and suggesting also that Hurley was the first to discover secrets that had previously been concealed to outsiders.[6]

As a title, *Pearls and Savages* was timely because the cinema industry in the 1920s was fascinated with the South Seas, coral reefs, cannibals, tropical islands, and native bodies. In fact, Sean Brawley and Chris Dixon observe how it was cinema in the twentieth century that gave shape to the identity of the South Seas region.[7] But the title also encapsulates succinctly and cleverly the contrasting aesthetics that Hurley and his generation found filmic and photogenic about coral islands: the romantic idea of a light-filled coast where pearls are found, and the equally romantic idea of the dark jungles where savages are found. Stephen Torre points out how the title *Pearls and Savages* reinforces a binary typical of the tropical island imaginary: "beauty and barbarity."[8] In other words, the dichotomous title promised audiences a great adventure for escape from the everyday, and an extravaganza of contrasts such as good and evil to engage their emotions and social values.

The same film was released in North America in 1923–24 as *The Lost Tribe*, and in Great Britain in 1924 to 1925 as *Pearls and Savages*. But the only version available for viewing today is a reconstruction created in 1979 by Keith Pardy at the National Film Archive of the National Library of Australia, in conjunction with Andrew McGuiness who performed the soundtrack, and researcher Andrew Pike.[9] The original screenings of the silent film were accompanied by lectures in which Hurley made sense of the material. So, when the film was re-created in 1979, intertitles were included as substitutions for Hurley's voice, some with direct quotes from

Hurley's diaries and from his book of the same name published in 1924. Robert Dixon characterizes the film from 1979 as "a new thing in the world, an object whose ontology reflects late twentieth-century assumptions about authorship and textuality."[10] In other words, the recreated film approximates but does not replicate the experience of original screenings. Nevertheless, it is this version that is the subject of discussion here, including its anomalies, but specifically the wildlife scenes of coral reefs at the Torres Strait and underwater scenes in still and moving images, captured using Hurley's aquarium on the Papuan islands of Dauko and Lolorua.

Hurley was confident that there was much he could improve about the first version of *Pearls and Savages* by augmenting, editing, and adding "sensational" new material such as moving images of tropical fish.[11] His decision was based on recognizing the kinds of moving imagery that were new to general audiences. It proved to be a good decision, because when *Pearls and Savages* was screened in London in 1924, reviewers made special note of seashore studies and also moving images of fish "abounding on the coral reefs."[12] Although photographed in black-and-white, the sight of fish swimming underwater was rare for a commercial film, and spectacular even though the effect of animal abundance was but a visual illusion shaped by the optical advantages of the glass-fronted aquarium. What is it like to watch the coral reef scenes and underwater photography in *Pearls and Savages*, and what was Hurley's photographic methodology?

As the camera surveys a coral reef exposed at low tide, the sight of the tide rising and falling over shallow coral pools is pacifying and hypnotic. There is a sense that this is a magical, dreamlike world. Although the audience sees the vision of the reef in black-and-white, the variety of coral forms is staggering, the expanse of the reef surprising, and the range of colors invoked by multiple black, gray and white tonal variations and reflections on mushrooming, blossoming corals, astonishing. Although black-and-white, the film enables us to understand a phenomenon that dominates Hurley's diaries: the kaleidoscopic colors of a tropical coral reef. An intertitle, with a quote from Hurley, tells viewers that they are about to see "the sublime wonders of the coral gardens." And sure enough, the reef appears as he said, like a garden bursting with shrubs and flowers of dazzling variety. A diary entry for September 22, 1922, describes Hurley's firsthand experience this way: "No gorgeous flower garden with its dazzle of spring blossoms growing in mass, could compare with this amazing marine garden of gorgeous madrepores, brain corals, soft corals & anemones. The formations were clustered in dense assemblage so that the sandy bottom was rarely seen."[13]

The first audiences of *Pearls and Savages* watched the sea washing over coral formations, rock pools, sea urchins, anemones, and starfish. A large clam snaps shut when someone pokes a stick at its mantel. An intertitle portrays a brittle star as a freak of nature for the way it regrows broken limbs. A black sea urchin bristles across a coral pool like an animated drawing by the symbolist artist Odilon Redon (1840–1916). Then the perspective moves to the underwater. The camera angle is fixed on a single viewpoint of a shoal of small reef fish swimming among corals with another larger species. What the audience doesn't see are the edges of an aquarium just outside the frame but excluded from the viewing field. Had the sides of the tank been visible, the entire illusion would have been shattered, and viewers would have realized that the ocean absorbing their attention was simply an artificial underwater.

The way the fish mass in still images figures 9.1 and 9.2, their abundance, and the visual impact of such varieties of shape and markings, stripes in particular, create a dazzling vision, a type of underwater orientalism that matches perfectly another section of the passage from Hurley's diary of September 22, 1922: "No jeweller if he spent his life could fashion in gold the supreme finesse with which the great master created the tiniest of these fishes. Colors that our purest of pigments looked dull alongside. Striped, banded, lined in fantastic patterns these fishes seemed to irradiate color—they seemed to glow & phosphoresce."[14]

In underwater scenes in *Pearls and Savages*, the fish get impossibly close to the camera. The focus of the camera is so close to the fish that if the footage had been filmed underwater, the fish would have been only millimeters from the lens. They are packed into the frame of the viewfinder, and swim mostly across the picture plane, parallel with viewers, not away from or toward them. They also appear to recoil from the sides of the frame, forced to return to the center by something out of the picture. And they seem agitated. The way they swim across the foreground plane indicates they occupy a shallow space, the closeness of fish to the viewer signals the presence of a glass barrier between the photographer and subject, and the way they recoil from the sides indicates the obstruction of the sides of the tank. Yet an intertitle inserted at the start of the underwater sequence (and placed in the film during reconstruction in 1979) tells view-

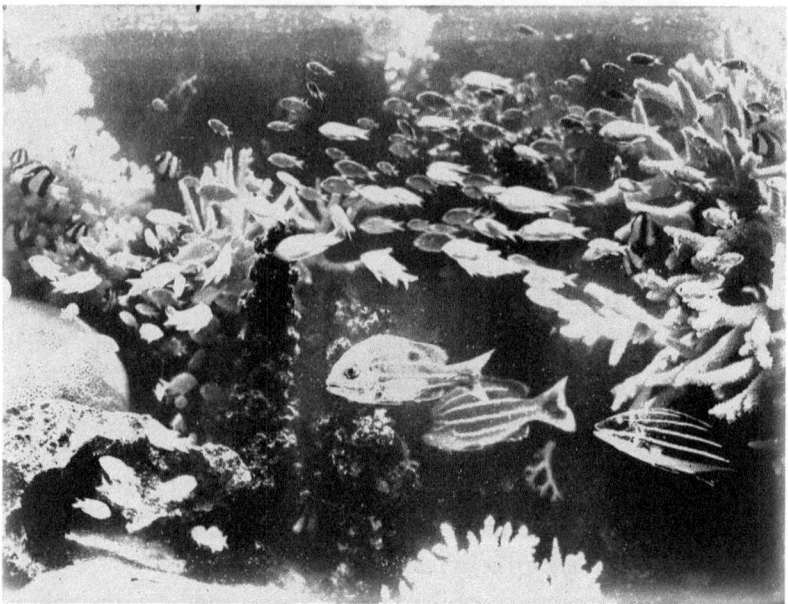

Fig. 9.2. Frank Hurley, *Coral Fish*, 1922. Black-and-white glass plate negative.
National Library of Australia, PIC FH/8962.

ers that "this wonderful underwater picture [was] taken from the porthole of a special diving tank." A "diving tank" implies a submersible, but Hurley never alluded to using a submersible. The intertitle wrongly suggests that Hurley was submerged while filming.

To the photograph-literate who are experienced readers of images, it will be obvious that the footage showing fish swimming underwater in *Pearls and Savages* could not have been taken underwater, or by a photographer occupying the same medium as the fish. The space from sea surface to sea floor is too shallow. It is a space impossible for the camera and cameraman to occupy. Figures 9.1 and 9.2 illustrate the point. The aquarium is invisible, but its presence, particularly its glass front, is no less palpable. By taking photographs directly in front of the tank, and by focusing on the center, the effect suggests a photographer who is on eye level with animals, and this produces a sense of authenticity in relation to the photographer being submerged in water. But at the same time there is a feeling of watching an event from a distance, and this is the effect of the aquarium's glass front. Glass is transparent, but it is no less an obstacle than other barriers to the immediacy of perception. Isobel Armstrong writes about how glass is paradoxical in the way it offers the sense of immediacy and spatial access while erecting a distinct barrier.[15]

In the early 1920s, though, the public was largely unpracticed at viewing underwater imagery and had little or no experience of the visual qualities of coral reef environments, of fish behavior in the wild, or of the materiality and optics of seawater. Public awareness of those things developed with the rise of tourism at the Great Barrier Reef in the 1930s and the emergence of new camera technologies, color photography, and new media channels such as television, all of which gave general audience new knowledge and new conceptual tools for understanding the underwater and its representation in film and photography.[16] Hurley, though, had traveled to the Great Barrier Reef in a period that can be described as a knowledge vacuum.

In 1923, the second release of *Pearls and Savages*, retitled *With the Headhunters of Unknown Papua*—Hurley referred to it as the "sequel"—was described as "the most marvelous motion picture ever screened."[17] The reaction of reviewers in 1923, in particular their excitement at the sight of "a submarine study of a shoal of fish sheltering in a cranny of the rocks," demonstrates the point that audiences accepted and did not question that they were viewing a true and real glimpse of wilderness rather than an artificial underwater.[18] Hurley knew that with careful lighting and the right perspective, and with the elimination of reflections off the glass front of the aquarium, there was scope to pretend that no barrier existed between photographer and the realm aquatic. Or was it that audiences suspended their disbelief in order to enjoy the experience of a viewpoint on the planet that was new? Had Hurley discovered a way, in that zone of visual slippage between the real and the artificial, to make his audiences forego logic for the sake of enjoyment?

If anyone had entertained the idea of sleight of hand, or what was then called "nature faking"—in which nature was staged or represented through artificial means—no one said anything; instead the popular press emphasized Hurley's originality. But as Susan G. Davis argues in her book about the history, sociology, and psychology of viewing undersea animal exhibits in public aquariums (and it applies also to domestic aquariums) it is always the case that "perfect seeing and visual intimacy are a trick."[19] One reason is the difficulty of getting so close to fish in the underwater, and another is the impossibility of transparent water offering clear vision.

In the underwater it is more usual to encounter dim and murky water in which bodies and objects are indistinct.

To return to the experience of watching the film, after audiences are shown scenes of fish swimming in underwater coral grottoes, *Pearls and Savages* focuses on a series of still photographs of fish, such as those in figures 9.1 and 9.2. They are odd, frozen doubles of the previous moving scenes and have a weird presence in the film. For one, in their stillness, they give the impression of being memorial photographs of the fish that were, just minutes before, shown moving and alive. But their placement is also strange because it forces viewers to shift their thoughts to something completely different from nature study: the reproducibility of photographs and how still and moving images of fish differ formally. They offer an incongruous deconstruction of photography itself, as if the idea is not just to depict wildlife scenes but also to expose and flaunt the tricks and powers of photographic processes. Comparing the representation of the movement of animals in photographs with the movement of animals in film was important to the history of photography and filmmaking, as Palle B. Petterson points out in relation to the nineteenth-century photographer Eadweard Muybridge (1830–1904).[20] But it sits strangely in the context of *Pearls and Savages*, in which the techniques of film and photography, including the use of an aquarium, were disguised in an effort to capture the illusion of authentic underwater space.

Much insightful commentary has been written about the social, commercial, and political contexts for the production and reception of *Pearls and Savages*, especially by Robert Dixon. But more can be said about Hurley's methodological approach to the acts of filming and photographing its scenes, including the problems of light and optics, the nature of visual apparatuses, and the experience of the eye perceiving bodies in water, all of which is detailed in Hurley's diaries. It is a rich record: both Hurley and McCulloch recorded for posterity the physical interactions they had with marine animals, as they trained moving, still, and stereoscopic cameras on the aquarium's glass front.

Before the fish were placed in the aquarium, Hurley and McCulloch crafted enchanting tableaux from the coral skeletons they had bleached in the sun. Manipulating the direction of sunlight was no easy task (fig. 9.3). Hurley wrote, "The aquarium has been set up on a convenient ledge facing East. The front tent entrance is laced over the glass front of the aquarium & the whole erected. The light filters through the water surface. This

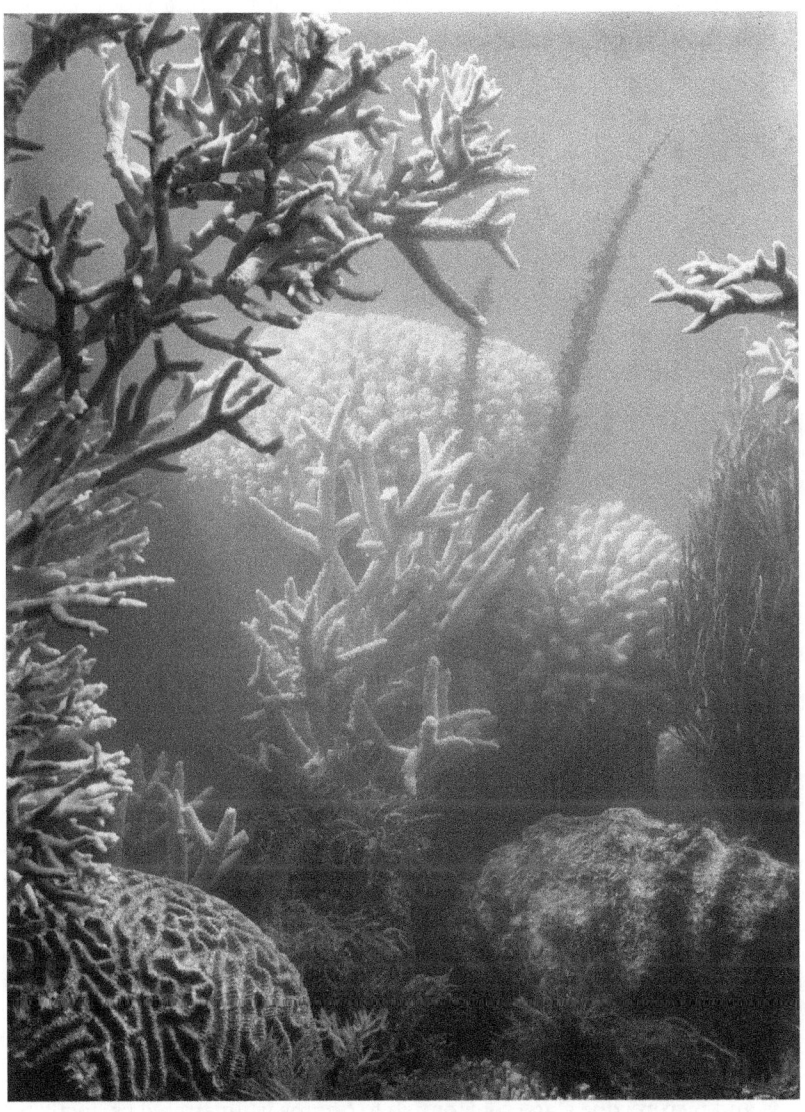

Fig. 9.3. Frank Hurley, underwater scene, Dauko Island or Lolorua Island, Papua, 1922. Courtesy of Australian Museum Archives, Frank Hurley Photographs AMS320_V05076.

method of lighting is similar to that which occurs when the coral & fishes are in their natural haunt."[21]

Hurley and McCulloch created an illusion of receding space using a Renaissance pictorial technique known as "repoussoir" whereby the scene was framed at the edges by coral branches that were placed in shadow to make them dark in color. This visual device encloses the scene but also has the effect of pushing the eye backward to a lighter middle ground. Repoussoir is a technique of distancing, of setting things away from the eyes so that scenes appear both contained and remote. At the same time, it heightens the sense of three-dimensional space. The scene Hurley and McCulloch created in the aquarium produced an effect rather like looking through a shop window or into a diorama.

Figure 9.3 illustrates what happens when vision is technologically enhanced by the magnification and transparency of glass, the clarity of purified water, the controlled play of light refracted through water, and the shallow space of an aquarium. As optical devices, aquariums are very similar to film sets and dioramas. Judith Hamera explains how they draw "on the perceptual logics and plots of the shop window, the theater, the panorama, and literal and fictive travels to become an amalgam of all of them."[22] As Hamera also details, the aquarium and diorama, like windows and travel, were examples of "spectatorial technologies, media, and positions" that in themselves helped to modernize spectators.[23] Marine life could be observed close up and on eye level, and this added to new knowledge, while at the same time the optics of water and glass, and the distortions of shape and movement created a spectacle to delight the mind by creating the aura of newness. Much the same as a diorama, an aquarium is a viewing device for marine life that produces in the viewer a sense of involvement and presence in what is otherwise an artificially organized scene. Unlike a diorama such as the Bahama Islands diorama in the Field Museum in Chicago, in which the animals displayed are taxidermied, though, an aquarium exhibits animals in motion.

The efforts Hurley and McCulloch went to in order to create mysterious and picturesque stage sets for fish to occupy is illustrated perfectly by figure 9.3. They tried to simulate the sights they had seen on the reefs, and the emotional impact of awe, as Hurley describes in the book *Pearls and Savages*: "You can walk from Dauko shore waist-deep through tepid waters of crystal sapphire, and tread the paths of silver sand amongst the coral beds. You hold your breath in wonder!"[24]

On Dauko Island, Hurley had grappled with Nature as a philosophical problem. After observing the variety of colors and forms among fish, anemones, clams, and corals, he asked, "For what purpose do they exist? He alone knows. Not even the beauty of His work is seen by many."[25] Hurley was perplexed by the purpose and point of so much beauty and variety when so few human eyes would ever witness it. Were coral reefs things-in-themselves, or things-for-him? If coral reefs are independent, self-existent entities requiring no human observer, then what is the status of Man? Does it mean that the beauty of nature is not for Man? The answer seemed wholly supernatural and beyond reason, which for Hurley put Nature in the realm of the other that is alien and external to humanity. The point is significant because it gave Hurley a sense that the nonhuman world of corals and tropical fish was something he was not part of. Allen Carlson names it "the mystery model [of nature] in which nature is alien, aloof, distant, and unknowable."[26] It is the very concept of nature that Timothy Morton wants to see "wither away" because not only is it romantic but it also creates a subject-object distinction that lies at the root of the environmental crisis. When human life is believed to transcend something called "nature," it leaves the environment vulnerable to delusions of human mastery.[27]

The image in figure 9.3, being so expressive of nature as a precious jewel external to human existence, was selected more than once for reproduction in magazines and the popular press, in which it was routine to represent wilderness as something alien to human life. The image was published in 1936 in an article by C. B. Christesen in *Walkabout* magazine titled "Roving the Coral Seas."[28] This staged, artificial scene was the kind of underwater view of the Great Barrier Reef that people wanted to see, and it was how Hurley wanted the Great Barrier Reef to be imagined. Stephen Torre explains Hurley's creative process and its impact on future generations this way: "The staging and replication of content so as to reflect what Hurley wanted to see in his subjects amounts to the substitution of a hyperreal, which then establishes itself in the discourse of the tropical island imaginary."[29]

When the image that appears in figure 9.3 was published in *Walkabout*, the caption described it as "A coral formation in a Barrier Reef pool." There was no mention of the intervention of an aquarium. Was the public likely to question whether the caption was telling the truth about the image, and whether the image was a true reflection of the caption? "Will not the caption become the most important component of the shot?" asked Walter Benjamin in 1931 in an article titled "A Small History of Photography."[30] Hurley's captions were vitally important for directing the imaginations of viewers to specific places—in this case, the geographical location of the Great Barrier Reef. But the relationship of captions and images can create plenty of confusion, a case in point being the title beneath a photograph of the Bahamas in *Mad Love* that reads "The Australian Great Barrier." Similarly, writers have more than once described Frank Hurley as a pioneer of underwater photography simply because Hurley insinuated as much through carefully worded captions, labels, and newspaper stories.[31]

The photograph seen in figure 9.3, taken in 1922, was also published in 1926 in an article Hurley wrote for the *Sun* newspaper titled "Neptune's Garden: Among the Coral of the Great Barrier."[32] The article stated that Hurley had just returned from the Torres Strait, "where he made a number of successful experiments in underwater photography."[33] In other words, Hurley had recycled an old image to fit a seemingly new story. But the story was the same one he had told in 1920, when he first claimed he had successfully photographed and filmed the submarine realm. When the story was circulated again in 1926, it became clear that Hurley's desire for recognition as a pioneer of underwater photography, and as an experienced deep-sea diver, was something he would never let go. He intended to make sure it remained part of the Frank Hurley myth—that of an extraordinary explorer for whom no region of the planet was too big an obstacle.

When fish were caught and released into the aquarium on the shores of Dauko and Lolorua Islands, Hurley's response was to say that "the effect was beautiful beyond words. They speedily accustomed themselves to their new environment & we could examine their evolutions & habits at ease."[34] To film the fish, he had to lie on his stomach. With the aquarium sitting on the ground, it was the only way to obtain a naturalistic, close-up view.[35] Hurley's intention with the close-up was to make the underwater scene appear immediate and knowable. So, whereas in 1921 he had peered over the side of a boat to look down on a coral reef, in 1922 he shifted that perspective to what Stephen Jay Gould refers to (in relation to aquarium photography and natural history) as the naturalistic, "edge-on view where a human observer sees marine life from within—that is, as if he were underwater with the creatures depicted, and therefore watching them at their own level."[36]

Not only Hurley but McCulloch, too, appreciated the eye-to-eye view, with McCulloch later arguing in an article titled "Fishes and the Movies" that "viewed from the side, as in an aquarium, fishes appear much more beautiful under water than when seen from above."[37] As far as McCulloch was concerned, because fish are never static, only moving pictures showing fish in motion were valuable for science. He was not bothered by the artificial nature of the aquarium because he knew that a tank was the only practical way in the 1920s to represent fish at ease in their natural medium.[38] Like Eadweard Muybridge, he wanted to capture the movement patterns of animals, but patience was needed to control where the fish swam. Unlike Muybridge, he and Hurley had ventured to the wild to take photographs of animals in their natural habitat. The English photographer, by contrast, was limited in mobility by the process that made him famous: a complicated photographic setup of multiple cameras that meant his work was confined to local sites such as zoos.[39] Hurley, though, was faced by a multitude of different problems, the most difficult of which was to bring every photographic consideration into play in difficult conditions in a makeshift film studio on the shores of a coral island. Capturing fish in motion depended on whether he could control the light and shadow conditions created by the outdoors (fig. 9.4).

Hurley and McCulloch wrote at length about eliminating reflections on the aquarium's glass. The machines and apparatus they used had to be invisible in the final film and photographs or Hurley would forfeit the entertainment and wonder value attached to viewers imagining that he was submerged while photographing. In this regard, his ambitions mirrored the aspirations of pioneer filmmaker Georges Méliès, whose production always kept invisible "the industry which produced the magic—rail, harness, gears, glass plates—and especially the cinema camera itself."[40]

Hurley and McCulloch positioned the tank so that it was lit from the sun through the top, but turned the front of the tank toward a mass of dark bushes to reduce the number of reflections. As McCulloch explained, the reflection of the movement of the handle of the cinema camera prevented them from capturing a seamless illusion of underwater space; it threw back to the viewer the processes of simulation, revealing the constructed rather than natural character of the film.[41] The circumstances were the same as those that Edward Eigen describes for the late nineteenth-century French zoologist Louis Fabre-Domergue, who also photographed marine life through an aquarium:

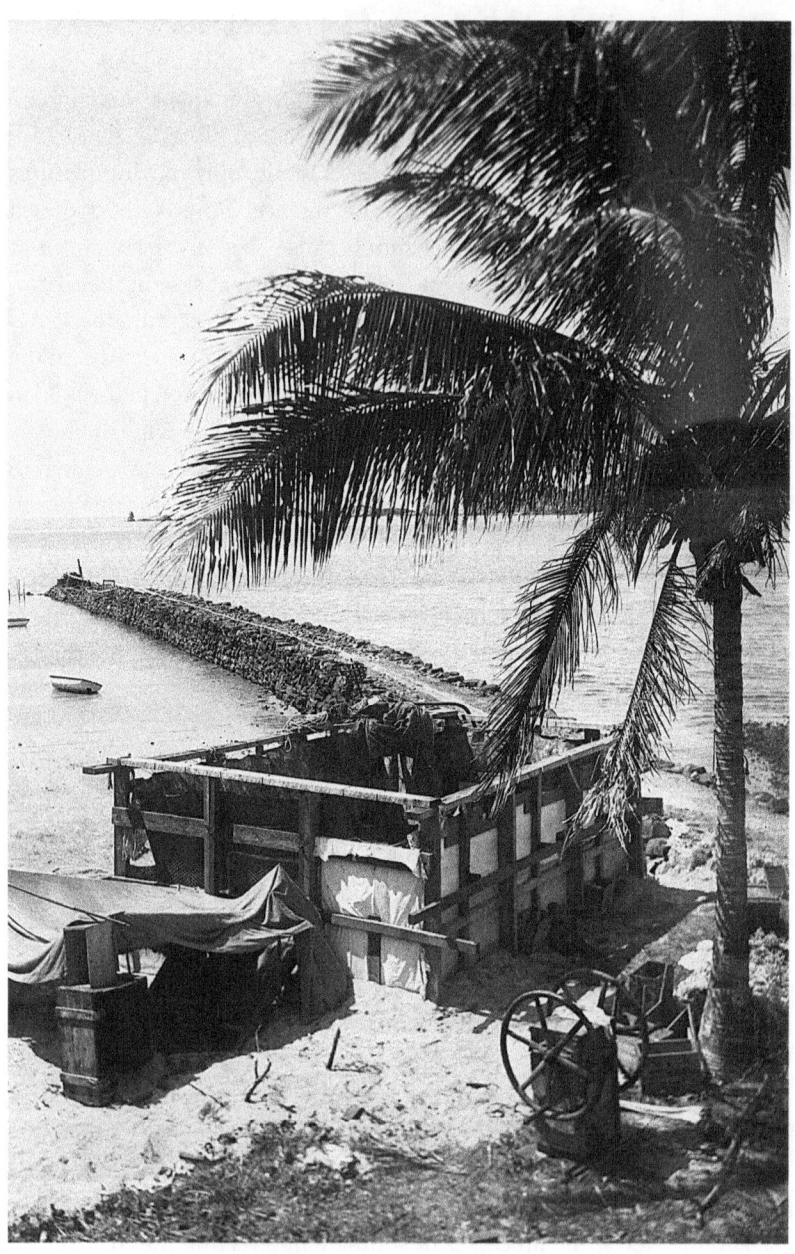

Fig. 9.4. "Capt Hurley's plant to photograph motion pictures,"
Dauko Island, 1922. Courtesy of Australian Museum Archives, Frank Hurley
Photographs AMS320/V4707.

Yet the camera-readiness of the scene equally attests to Fabre-Domergue's rational scrutiny of each of its elements. He analysed the aquarium's geometry, the qualities of each of its surfaces, and a variety of lighting effects. He learned to do so from the faults in construction that the camera itself revealed, including glare from the camera's lights and unworkable angles of vision. In a new evolution of the Albertian picture window, the aquarium became a visual passage to an interior space; the medium it enclosed, however, did not readily transmit a focused, undistorted image of the subjects suspended in it.[42]

Hurley, like Fabre-Domergue, struggled with the camera's optics in his effort to achieve a naturalistic underwater scene, and like Fabre-Domergue was forced to partition the space of the tank into two. The smaller enclosure ensured the fish could not hide; it pushed them to the front for maximum visibility. He and McCulloch also devised pipes to transport water from the shoreline, whereas prior to this they had made the mistake of filling the tank at the edge of the beach and carrying it to the right location.[43]

After Hurley and McCulloch worked out how to keep the fish to the front of the aquarium and within the camera's frame, their diaries became quite detailed about the appearances and behaviors of fish, particularly how skin patterns and colors changed according to light conditions. Hurley noticed the impact of the camera on the fish, how they reacted to their new artificial environment, and through observation came to understand the integral nature of aesthetics to their identity:

> The aquarium with the vertical sun streaming down through its surface was one of the most beautiful things I have seen. . . . The coloration of the fishes was beyond comprehension. The small shoals emerald green fading to a light green under the belly. When the sun shone through them they seemed transparent like opal. The larger fishes bright yellow with longitudinal blue lines. In one corner a small group of bright red fishes hid away in a coral clump. In another clump a number darted in & out amongst the coral white & black banded. I had a great photographic day exposing over two dozen plates & 500 feet film.[44]

Through attentiveness, Hurley also came to understand the theory of animal camouflage that was current at the time, a principle known as "obliterative coloration," a term coined by the American naturalist and artist

Abbott H. Thayer (1849–1921). "Thayer's Law," as his theory was known, hypothesized that in the right lighting conditions, gradations of light and shade on the bodies of animals, namely dark topsides and light undersides, produce an illusion of transparency, making the animals seemingly disappear in plain sight.[45]

McCulloch described one species, *Chromis*, as particularly photogenic. Although photographed in black-and-white, figures 9.1 and 9.2 offer views of the small *chromis* fish in the aquarium. Not only was the *chromis* "opalescent green" and spectacular to look at, the species, wrote McCulloch, also made "particularly good movie models" as they moved between the corals and obliged by feeding directly in front of the aquarium's window, "swimming in and out, continually in motion, and therefore a star performer for the cinema."[46] McCulloch's ironic self-reflexivity about the spectacle that he had helped invent, and the choice to depict the personas of fish as film stars, illustrates the point that McCulloch the scientist, like Hurley the artist, observed no strict differences between science and art, or between pedagogy and entertainment.

Three "genres" overlap in the films and photographs of tropical fish underwater that Hurley created in Papua—the pictorial (that is, dazzling forms of corals and fish), the documentary (an objective record), and the revelatory (the never-before-seen underwater photograph of tropical coral reefs). These were images that could be read, and were read, across the contexts of art, science, and popular culture. And as Jennifer Tucker explains, "It was the mass distribution of photography through industrialization that softened borders between art and science through the popularization of scientific images."[47]

The *Illustrated London News* wrote that *Pearls and Savages* showed that Hurley was "A Picture Sorcerer in Papua."[48] He moved among "headhunters," recording their lives and customs. He photographed coral reefs and tropical fish in the underwater realm. Unlimited mobility and exceptional courage were signs of his magical powers. Traveling in lands in which magic and sorcery were part of everyday life, Hurley had shown that the white man had magic of his own. Or so the newspapers insinuated. It was all part of the melodrama and theatrics that accompanied the publicity surrounding *Pearls and Savages*.

On October 29, 1922, sailing back to Thursday Island in the Torres Strait at the conclusion of his expedition to Papua with Allan McCulloch, Hurley shone a light at night on the sea over a coral garden and watched as fish gathered around it like moths to light. At the end of his trip he determined to return again, vowing that "the most I can say is that in the near future I will concentrate on the Great Barrier Reef: for I can think of nothing more beautiful in creation."[49] And return he did, in 1926, to film a narrative South Seas drama titled *Hound of the Deep* (also released as *Pearl of the South Seas*), a film written, photographed, and produced by Hurley, and described in 1926 as having perfectly captured "the adventurous and romantic atmosphere of the Southern Seas and transposed it on to the screen in a turbulent romance that rises to the highest peaks of dramatic intensity."[50] By 1926, educational and travelogue films like *Pearls and Savages*, films that had thrived in the first two decades of the twentieth century, were in less demand, as were films directed toward colonial education about imperialist geographies.[51]

Hound of the Deep explores action and romance as a young man from England travels to the Torres Strait in search of a pearl so large it promised to outrival all others in size. In underwater scenes, the young man collects shells while a villain is trapped by the foot in a giant clam, sights that were major draw cards for cinema audiences excited about witnessing the floor of the sea. In the *Sun* newspaper in 1926, Hurley claimed that to secure the deep-sea scenes, he used a watertight camera operated electrically.[52] Immediately this conjured for readers a vision of Hurley, the diver, descending to the bottom of the Torres Strait. Watching the underwater scenes is atmospheric and mesmerizing. But they were filmed through the glass wall of a large tank, the device being most obvious when an octopus stretches its way across the glass front. The deep-sea scenes were not filmed in the ocean. Yet this is the same *Sun* article in which Hurley compared himself to J. E. Williamson and claimed that he was more mobile in the underwater than Williamson, that he was a practiced diver and an experienced underwater photographer, and that he had photographed the floor of the sea at the Torres Strait.

There are many things about Frank Hurley's life that are ambiguous and uncertain, among them his claim to be a practiced diver. But the wonder and astonishment of the diver was something Hurley had long thought about. The symbolism of diving and the heroics of diving were instilled in his psyche well before the first and second expeditions to the tropical north of Australia. In fact, any effort to portray Hurley's efforts to represent the underwater would be incomplete without an account of the significance of diving to his work and self-image. He was fascinated by all aspects of diving, and had predetermined before the outset of the first expedition to the Torres Strait in 1921 that he would devote a special scene of *Pearls and Savages* to the eerie-looking men who dressed in suits and helmets and descended to the seafloor to collect pearl shells in a region commonly described by Westerners as one at the world's outer margins.

Although Hurley did photograph pearl divers on the first expedition to the Torres Strait in 1921, and included a brief scene in *Pearls and Savages*, the footage shows divers in suits above water, and on deck. In 1926, though, a few Hurley photographs started to emerge in the popular press showing divers on the seafloor collecting pearl shells. But the setting is an artificial underwater, indeed it is the same tank used for *Hound of the Deep*. The still photographs of underwater divers were distributed by photo agencies to Australian and international audiences. They represent a second body of underwater photography in Hurley's repertoire of images, and are different from the collection that shows fish and corals underwater because these images include the human figure.

CHAPTER 10

Hurley and the
Torres Strait Diver

The idea of diving the underwater was important for Frank Hurley's self-image. His preoccupation with diving may have originated from reading engineering and mechanical magazines, particularly *Popular Mechanics* and *Popular Science*, in which people who undertook diving for scientific discovery, industrial advancement, and undersea exploration were depicted as modern heroes. Whatever the source, by 1915 diving was part of the Hurley myth. A defining incident took place on November 2, 1915, during the Shackleton Expedition. Hurley had dived underwater to save a collection of negatives and photographic instruments from the sinking hull of the shipwrecked *Endurance*: "I hacked through the thick walls of the refrigerator to retrieve the negatives stored therein. They were locked beneath four feet of mushy ice and, by stripping to the waist and diving under, I hauled them out."[1]

As the years passed, the story grew more elaborate. In 1924, for example, when retelling the incident in the *New York Times*, the depth of the dive to rescue photographic gear from the sinking *Endurance* in Antarctica, expanded to "fifteen feet of pea-soupy ice."[2] By extending the depth from four feet to fifteen, which is the difference between standing in water and being out of depth, the story, and Hurley, became more impressive.

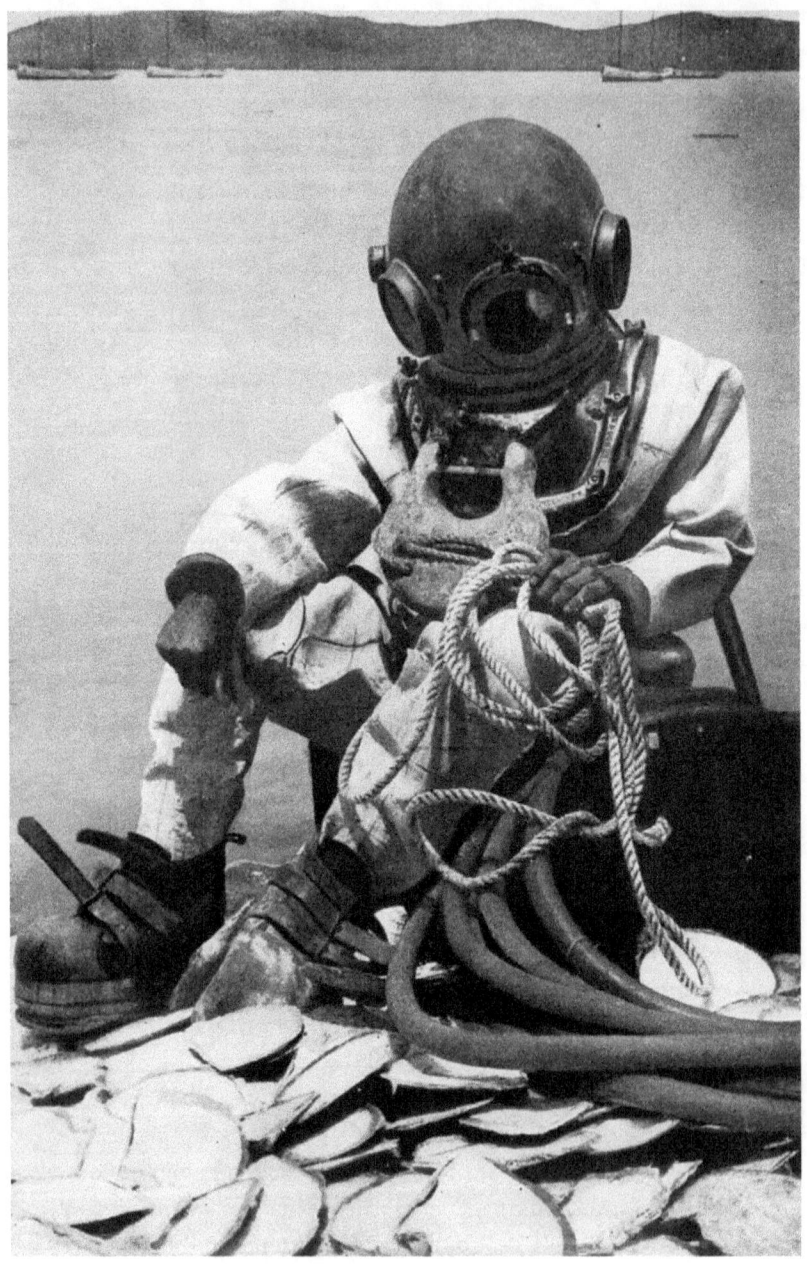

Fig. 10.1. Frank Hurley, "A diver rigged ready for descent," ca. 1922.
Australian Museum Magazine, January 1924, frontispiece.

From 1921 to 1936, Hurley was commissioned by the Sydney *Sun* news-paper to write diving stories. The *Sun*, a populist tabloid, wanted exciting articles, and in September of 1921, just after Hurley's return from the first expedition to the Great Barrier Reef, the Torres Strait, and Papua, it published "Beneath the Waves," an account by Hurley of diving in "strange grotesque life" among a "forest of amazing foliage" in Sydney Harbour. Hurley explained how he had descended to the floor of the harbor in a self-contained suit in the company of Norman B. Friend and "Diver Carr," two of the harbor's earliest underwater explorers. Hurley began the account by posing one of the key questions of modernity: "What and where will the people turn for adventure" when the planet's virgin spaces and wilderness are lost to the expansion of cities and populations? Already space was shrinking, and the blank areas on the imperial map were disappearing, but Hurley was confident: "The spirit of adventure will never be assuaged, and is as necessary to mankind as the wild free air itself. Then there still remains the ocean."[3]

The article was mostly a description of dress diving in Sydney Harbour off Shark Island. Hurley wrote how he had taken a submarine cine camera with him, describing it as "a cinematograph with exceptionally large aperture lenses sealed in a heavy casing and actuated by a small pneumatic motor driven by compressed air." A description of his descent to the seafloor followed:

The violent bubbling of escaping air sounded in my ears, and I committed myself to the waters of mystery and doubts.

Surrounded by a luminous green aura, spangled with scintillating showers of shell grit and foreign matter, "spot-lighted" by sunbeam radiance, dancing like playing rays from a cinema projector—I went down and down and down. Small fishes glided through the opalescent glooms, circled within hand reach, manoeuvring with ineffable grace then glorified by shimmering beams, flashed silver and went whence they came. Beneath was a hazy shadowland of waving foliage and strange shapes: rocks ribbed with crevices encrusted with shellfish clumped with growths and bristling sea urchins.[4]

The description of light entering water is memorable. To conceive of the underwater as a cinematic space, a submarine shadowland where the sun's rays shine down the way a film projector throws light rays into a picture theater, is Hurley at his most poetic. He conceptualized the ocean not only as a giant aquarium but also as a giant cinema for projecting dreams and fictions as well as a space to colonize and play in. As the text progresses, though, it is difficult to reconcile the contradictions between reason and imagination, the absurd and the logical, romance and scientific data. To a contemporary reader, at least, Hurley indulges in imagery that is simply not believable.

On the one hand, the reader is asked to consider the facts of life for a deep-sea diver battling currents and other adversities, much as in a research essay. But on the other, some passages expect the reader to suspend disbelief and imagine Hurley sitting "in a quiet bower surrounded by all this exalted grandeur."[5] Hurley's descriptions of the "luminous green aura" of the submarine realm are evocative, but we are left wondering if the phrase wasn't a stock standard description of underwater optics. Norman Friend, who often published his own accounts of diving in Sydney Harbour, also noted the "luminous green aura" in his writing.[6] Only with visual evidence of Hurley's dive with an underwater camera in Sydney Harbour can we corroborate the story that appeared in 1921 in the *Sun*. Otherwise this part of Hurley's life, detailing underwater diving and cinematography, will remain unresolved.

In 1926, Hurley published in the *Sun* two more stories about diving adventures. One described a personal experience of free diving in the Torres Strait. Hurley was at the height of his career when "Neptune's Garden: Among the Coral of the Great Barrier" came out. In the context of Hurley's career, this story is unique for the way it adopts a first-person viewpoint of coral reefs underwater. Nothing like it had appeared in Hurley's news stories before, or in diary entries of the two expeditions to the Great Barrier Reef, the Torres Strait, and Papua in which wading is mentioned but not diving or swimming. The *Sun* article begins with the reader being invited to join Hurley "for an undersea ramble in Neptune's Garden" in the company also of "a half-blooded Samoan, a fearless and magnificent bareskin diver and swimmer," a man called "Leo" who was Hurley's coxswain:

I rise for breath, and dive again to glide over fernlike fields of dark brown weeds and clumps of green ribbon grass, that trends and waves in the invisible current. Shoals of tiny fish scurry and flash like meteor showers over bushes of pink corals, shrubs of red, and antlers of blue. It is like a sublime kaleidoscope. . . . Here lies a submerged world that outrivals the glory of the starlit firmament. . . . Once more I rise to the surface, and Leo informs me he has found a giant clam. I dive and follow him—what an elegant swimmer he is, and how graceful his movements under water! We reach the grim yet gorgeously colored shell-fish. The shells are at least three feet across. Leo swims close, and prods it with his spear. Instantly the huge vice-like shells "clop" together. To be gripped in such a monster would mean a broken limb, and unless assistance was at hand to free the captive he must be drowned.[7]

The way Hurley described this extraordinary diving moment suggests that he was completely at home in the underwater, swimming effortlessly and in rhythm with nature, rising, surfacing, and diving with similar ease and grace as his diving partner, and able, seemingly, to live comfortably in both air and sea. But something is awry, because illustrating the *Sun* article is the image in figure 9.3, an aquarium photograph taken four years earlier during the camping trip with Allan McCulloch on Dauko and Lolorua Islands. Used as an illustration for "Neptune's Garden" in 1926, the photograph insinuates itself as a picture taken during Hurley's dive with the Torres Strait Islander free diver. The *Sun* story also recycled descriptions of coral reefs that were published five years earlier, including the sight of a "grim yet gorgeous" giant clam, "meteor showers" of luminescent fish, the undersea as "a furious battlefield," and the "sublime kaleidoscope" of the reef's forms and colors. Hurley had simply retrieved them from a storehouse of phrases and descriptions filed, mentally at least, under the subject headings of "coral reefs" and "the underwater."

But taken at face value, Hurley's article in 1926 was a thrilling story that gave city readers vivid images of the South Seas and confirmed long-standing fantasies of life on coral islands affording freedom of mind and body, as well as escape from the metropolis, modernity, and the ills of postwar Europe. It also fitted well with the primitivist turn in Western thinking at the beginning of the twentieth century, especially in artistic circles, when the region of the Torres Strait, Papua, and New Guinea was attractive to artists wanting to obtain insights to people said to be still living in the "stone-age" who were also believed to live close to nature. Consequently, in films and photographs, in paintings, and in texts of the time, the concept of "native society" was represented as a phenomenon of slow, "traditional time."[8]

The surrealist poet Paul Éluard (1895–1952), for example, in 1924 traveled from Paris on a long sea voyage that included New Zealand, Australia, the length of the Great Barrier Reef, the Torres Strait, Papua New Guinea, and the Trobriand Islands. Robert McNab explains how Éluard entered a "territory of great imaginative and cultural value" to Western art.[9] In the Dutch colony of New Guinea he witnessed the realities of the impact of colonialism on Indigenous peoples. Later he amassed a collection of masks and tribal artifacts from the region. His journey intrigued and impressed fellow artist André Breton, who, in the same year as Éluard's return from the region, wrote the first Surrealist manifesto arguing against the rationalism of modern society and "the incurable mania of wanting to make the unknown known, classifiable," calling instead for a new era of exploration, one of the mind and the imagination and everything that "sinks back below the surface."[10] In 1937, the idea of metaphorically sinking below the surface of the mind to the unconscious, of sinking geographically below Europe to the Antipodes, and of sinking below the surface to the sea to the underwater were all evoked by Breton when he published in *Mad Love* the previously discussed submarine photograph of a Bahamian coral reef that he titled "The Australian Great Barrier."

Primitivism also infused Hurley's imagery and writing about the sea and the underwater. In the 1920s, primitivism was associated with the avant-garde, but, as Robert Dixon explains in connection with Hurley, it was also a "modern commodity" sought after by the mass entertainment industry that supported and encouraged films like *Pearls and Savages*.[11] In keeping with primitivism, passages of the *Sun* article "Neptune's Garden" evoke Hurley as a free child of nature and a modern savage exploring the submerged coral sea. He embellished the idea of personal, social alienation from the modern city to strengthen the intensity of the story. But separations of race, rather than a utopian unification, overshadow the text when Hurley characterizes himself as a white man returning temporarily to a primitive past and to conditions his own race had long left behind. He wrote how, in preparation for the dive, he had colored his white skin black to fit in with the savage environment of the Torres Strait: "Could you but peep at us it would be difficult to pick me out from my native crew. I have blackened up! The reason is to disguise myself as a native, so that when swimming, prowling sharks may not single me out in preference to the others."[12]

The passage evokes stories by Rudyard Kipling. In "How the Leopard got his Spots," in *Just So Stories* (1902), a man described as a greyish-colored Ethiopian changes his skin color to blackish to better match his forest background, protect himself from enemies, and empower himself as a hunter. Kipling's story about blackness and camouflage influenced the popular imagination and helped construct race by naturalizing the idea that, in the primitive wilds, it is biologically true that being dark in skin is better than being light. In the first decades of the twentieth century, relations between human beings and their environments were shaped by this kind of race-thinking, and Paul Gilroy notes the reductive nature of presuppositions based in social Darwinism and the principle of survival of the fittest.[13] Following the First World War, the concept of a latent primitive masculinity in white races became prevalent in social theory.[14] It was applied also to warfare in the Second World War, when soldiers fighting in northern Australia, Papua, and New Guinea were taught that "being too dark is generally safer than being too light."[15] Frank Hurley's *Sun* article was therefore of its time in relation to viewpoints on skin color, nature, and race. In 1926, his story about blackening up to stay invisible to sharks would have entertained the *Sun's* readers and reaffirmed racial stereotypes and ideologies of the type that J. E. Williamson's films had also promoted.

Free diving and dress diving were topics that people in the 1920s were eager to read about, but dress diving for pearls was by far the most popular. Free diving, while intriguing, was of more interest to anthropologists and travelers who specialized in observing and studying the customs, habits, and skills of Islander peoples. Free diving was diving in the traditional sense, and many Western travelers and naturalists gained their first insight into the underwater by studying free divers at the Great Barrier Reef, the Torres Strait, Papua, the Bahamas, and the Caribbean. Louis Boutan, as previously mentioned, dived naked with pearl divers in the Torres Strait before becoming one of the first successful photographers of the underwater.

The history of free diving was fascinating, but the story of dress diving for pearls was both romantic and political. Political, because at the Torres Strait in the late nineteenth and early twentieth centuries, pearl diving was, to many, a story about successful colonial expansion, whereas to critics of colonization, and to those affected, it was a story about exploitation. Thomas J. McMahon, for example, wrote in 1920 in the *Geographical Review*, a journal published by the American Geographical Society, how the history of pearl diving at the Torres Strait was a story about Islander peoples in slavery. Men, he said, were "seized and kept almost constantly at work as divers; the women were scattered; separation from home and children ensued."[16] Western and Eastern markets could not get enough pearls, mother of pearl shell, bêche-de-mer (an animal also known as "sea cucumber" and trepan, and a delicacy when dried), trochus shell, sponges, and fish. The bêche-de-mer and pearling industries took the young men of Torres Strait villages away to work on pearling luggers; it was well known that Torres Strait Islanders were superior in underwater endurance.[17] But, as in other parts of the world, including the Bahamas, the skills of divers were exploited through cheap labor for commercial businesses. Not only were people misused but undersea marine resources were so exploited that by 1890 the riches of pearl shell beds in the Torres Strait were gone forever, according to the scientist Charles Hedley, who published an article in 1924 on the subject for the *Australian Museum Magazine*, illustrated with photographs of pearl divers in the Torres Strait taken in 1921 by Frank Hurley (see fig. 10.1).[18]

Yet pearl diving was romantic because the idea of a diver slipping beneath the surface of the sea, descending to the sea floor, and returning with treasures was shrouded in legend. In poetry, philosophy, and religion, pearl diving was a metaphor in its own right for discovery, enlightenment, and transcendence.[19] The first time Hurley wrote about pearl diving in the Torres Strait was 1921: "The quest of sunken treasure holds no stronger lure than the search for pearls. Fortune and her co-partner death ever collaborate to spirit the diver to the inevitable brink; groping through the slime, stumbling across the dread sea floor, staggering ahead to win her prize. Pearls are the reward of hardship and suffering—they are indeed emblems of sorrow."[20]

Hurley was attracted by the idea that the pearl diver was willing to go to the ends of the earth for treasure. He also knew that divers worked in peril in what was referred to in his time as a world of "cold, silence, and blackness."[21] He was shocked by the size of the graveyard on Thursday Island where pearl divers who had been killed in diving accidents were buried. However, by the time Hurley arrived in the Torres Strait, most commercial dress divers were Japanese and Malaysian, and the industry of pearl fishing, for pearls and shells, was coming to a close, the oyster beds being completely overfished.[22] Hurley nevertheless planned to make a study of dress diving. What is it like to breathe air underwater through a metal helmet, to walk along the bottom of the sea in a canvas suit with weighted boots, and get pulled along by a rope from a pearling lugger? These were some of the questions that Frank Hurley had in mind when he visited the Torres Strait in 1921.

Enthralled by the adventure and danger of dress diving, he devoted considerable time and effort to researching the operations firsthand. Hurley's diaries describe the risky method of towing workers along the sea floor as they breathed compressed air through hoses.[23] Diving for pearls took place in depths ranging from 10 to 45 fathoms (60 to 270 feet), but in the deepest water a diver could stay under only four minutes—longer than that and they risked paralysis. Hurley weighed the diver's equipment to put the life of a diver in perspective.[24] He learned that a diver regulates his height above the bottom of the ocean by opening and closing a valve on the helmet, allowing air to flow to the diving suit. Too much air is dangerous because "the diver shoots up to the surface."[25] From the deck Hurley photographed the suited men as they descended into the water, and again as they reemerged. And he published what he learned in two articles for the *Sun*.

But when the *Sun* stories were published in 1921, Hurley claimed that he too had worn a diving suit and descended to the bottom of the Torres Strait, although no mention of the same appears in the diaries. One article, titled "With the Pearlers: Land of Topsy-Turvydom," is not illustrated but contains vivid descriptions of the underwater, of darting fish, opal green ambience, and swaying seaweed. Hurley portrayed the bottom of the Torres Strait as an unexplored "prehistoric wilderness" in which he was "overawed by the unreality of this sublime world."[26] Similar descriptions would later appear in the story about diving in Sydney Harbour.[27] "With the Pearlers" is an interesting, if somewhat textbook history of pearl diving in the Torres Straits, and of the origin of pearls as biological forms and commodities. Hurley made the philosophical observation that a pearl is perfection born of its own inward struggle and conflict. But many passages are fantastical and comical and contradict the facts and dangers of diving that are detailed in diary entries. In one passage, for instance, Hurley himself "slid down sunbeams" to the floor of the Torres Strait, then, "like a submerged cork let go, turned a somersault and sped to the surface head down."[28]

To write arresting stories for the *Sun* and produce remarkable photographs of Torres Strait pearl divers, Frank Hurley knew it was important that the public imagined that he had followed the "pearlers" to the seafloor. In 1926, he produced a body of photographic work showing a pearl diver collecting shells in a string basket at the bottom of the Torres Strait (figs. 10.2 and 10.3). From 1926 through to 1950 photographs of the same diver, shown at slightly different angles, were reproduced in Hurley's stories.

Figures 10.2 and 10.3 show a pearl diver at work underwater on the seafloor. But the photographs were not taken in situ. This much can be deduced from the images themselves: the closeness of the photographer to the diver, the camera's position slightly below the subject, the stillness of water, the clarity of vision, and the faint suggestion in the background of the edges of a tank.

The images, though, are in themselves intriguingly ambiguous, which makes the subject of their genesis interesting. Each one shows the same diver in the same diving suit and helmet with the same basket in which pearl shells are being placed. However, as figures 10.2 and 10.3 demonstrate, some prints are cropped more tightly, the diver's hands and arms are in different positions, there are fish in some prints but not in others, the diver's head is at slightly different angles, and in figure 10.3 the diver can be seen sitting on the floor, almost cross-legged, a position that is disguised in figure 10.2 by the diver's left arm. For the image in 10.2 to be true to life, the sitting position had to be hidden because pearl divers did not sit on the ocean floor but instead plodded around with lead boots, bending with great difficulty to collect shells. The fact that the diver in figure 10.3 is sitting on the floor raises doubts as to whether he is in water at all, unless he is heavily weighted.

As a visual image, figure 10.3 bears little resemblance to the textual image of deep sea divers at work that Hurley crafted in words: "One hundred and twenty feet below the surface the scene is eerie and unforgettable. A hobgoblin army moves slowly, as if overwhelmed by a sluggish weariness, through the green gloom, groping its way through trailing weeds and growths. From each grotesque unit, streams of silvery bubbles dance upward to the bright ceiling of flickering green sunbeams. . . . As they advance, the divers bend to glean their prizes from the gardens of the deep."[29]

The imaginative power of this text, which evokes deep space, is in contrast to the awkward, close-up images in figures 10.2 and 10.3. Moreover, the text copies the grand, epic scope of literary and visual images of wilderness landscapes. The viewpoint is panoramic. It describes the seabed from the perspective of a lone diver who is apart from the group, seeing

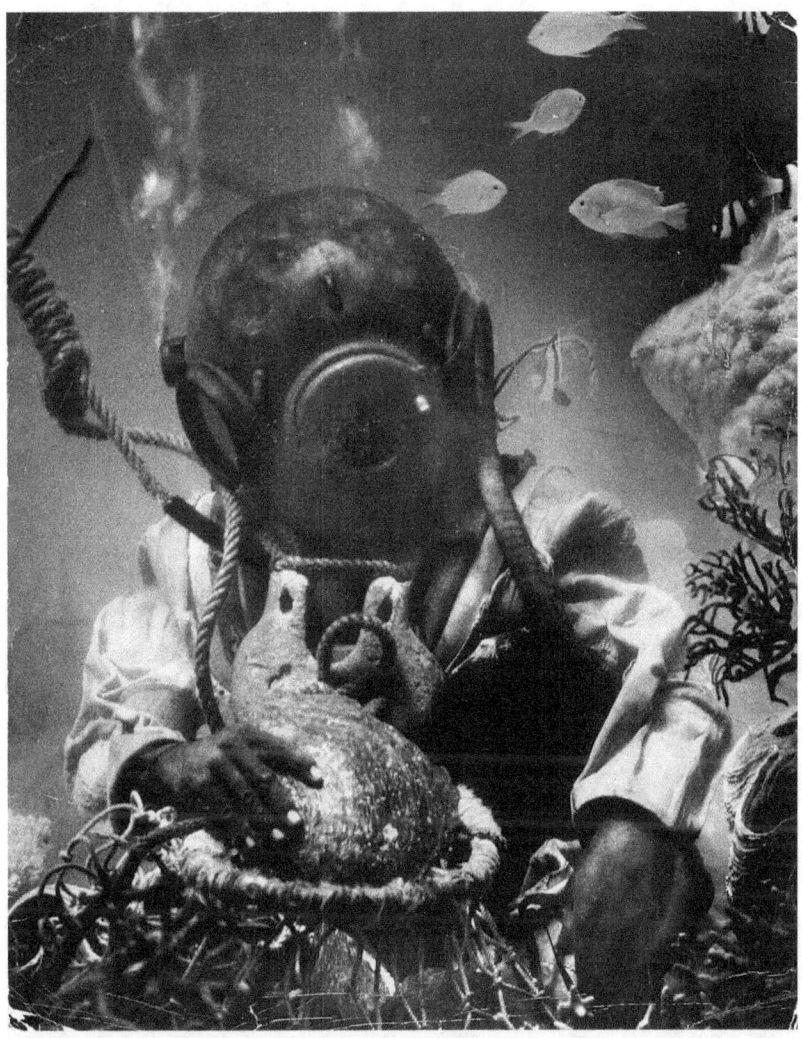

Fig. 10.2. Frank Hurley, "Pearl diver collecting shells from the beds of Torres Strait," 1926. Composite print. National Library of Australia, PIC/14197/120.

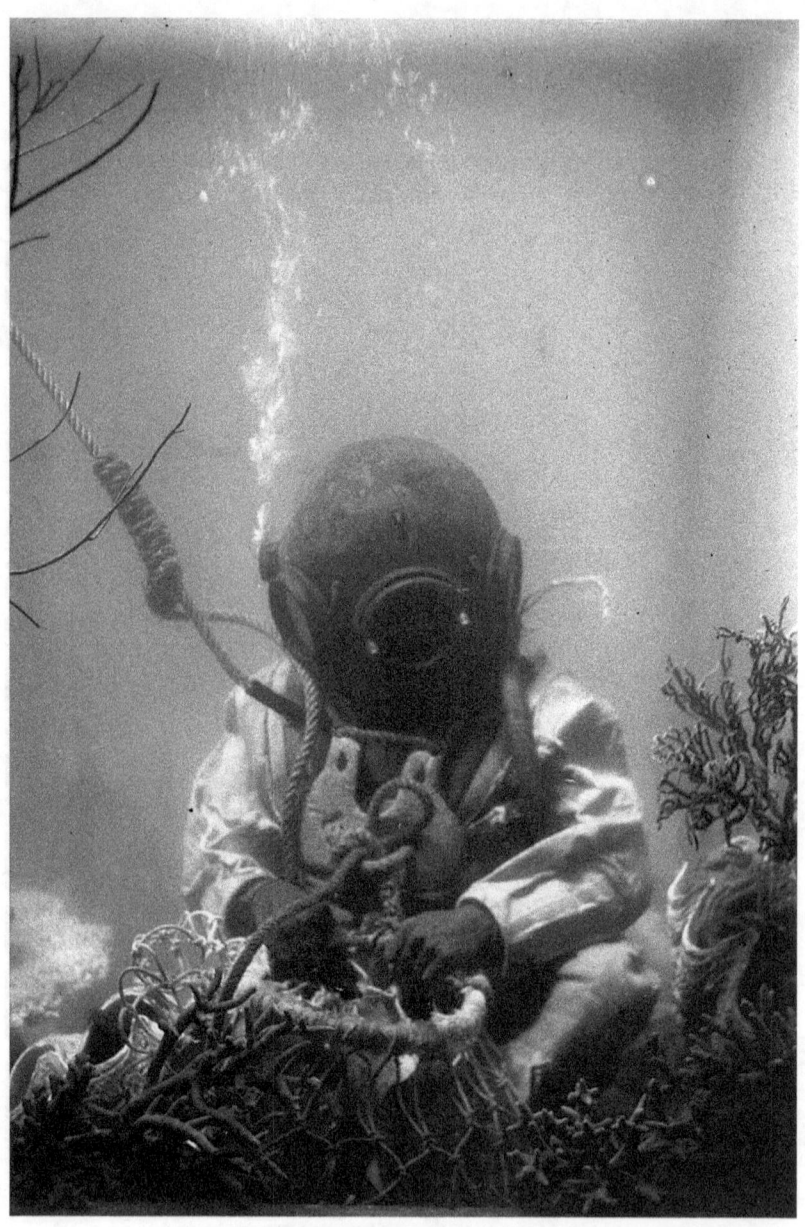

Fig. 10.3. Frank Hurley, Pearl Diver, 1926. Courtesy of Australian Museum Archives, Frank Hurley Photographs AMS320/V5071.

the rest of the party with impossible clarity and depth of field even though they are spread out over a hundred feet in the distance. In fact, the image seems to mimic the scene depicted in *Underwater Landscapes of Crespo Island*, an engraving by Édouard Riou and Alphonse de Neuville that illustrates Jules Verne's *Twenty Thousand Leagues under the Sea*. The illustration shows pearl divers trudging across the ocean floor in self-contained suits surrounded by the vastness of the organic underwater where canopies of kelp reach through space and where jellyfish float and undulate.

Because Hurley's written description of pearl divers working at a great depth on the ocean bed bears such a close resemblance to the illustration in Verne's book, we are reminded of the scope and variety of visual and textual images of pearl divers in popular, artistic, and scientific contexts that were available for Hurley to draw on to create his own stories and images. Among them we need to count Stuart Paton's *20,000 Leagues under the Sea* (1916), the movie for which the filming of underwater scenes of divers trudging through currents on the ocean floor was directed by J. E. Williamson. Through Hurley's sublime rendering of the scene we are also reminded of how modern technologies and the enchantment with speed and mobility had, as Alain Roger explains, "obliged" people to inhabit "new landscapes, underground, underwater, in the air."[30] To be truly modern was to have the capacity to explore and claim any space at all on the planet.

Hurley's written text, quoted above, invokes Jules Verne, and the image it conjures is of an underwater view imagined within the pictorial conventions of the fine art landscape tradition also observable in the illustration *Underwater Landscapes of Crespo Island*. The text exalts solitude and danger. Yet the visual images in figures 10.2 and 10.3 have more in common with early twentieth-century lithographic poster illustrations, comics, and the front covers of popular culture journals, in particular, *Boys Own*, *Popular Mechanics*, and *Popular Science*, in which divers, including pearl divers, are shown close to the picture plane, in a vertical format, low in angle and large in scale. About comic book representations of underwater divers and landscapes published in the early twentieth century, Laurence Le Dû-Blayo explains that they typically give a magnified view of the underwater "to immerse the reader in the underwater world."[31] The divers in Hurley's photographs are presented as alien figures, human-machine hybrids, absorbed in the strange space of the undersea. It is not ecology and landscape that matter in these popular culture images of divers, or in Hurley's photographs of divers, but the theatrics of the figure and the immediacy of the image. They reflect the increasing stress on visual impact in modern mass media arts. Hurley's staged photographs of deep-sea divers are projections of an imagination fed by multiple influences. As images engaging a public, they reflected back the idea that the wilderness of the undersea was fast becoming tamed and domesticated through modern technology and by modern explorers and adventurers. They also inferred that Hurley was an agent of that change.

The photographic print shown in figure 10.2, which is part of the collection of the National Library of Australia, is inscribed on the back by Hurley: "Pearl diver collecting shell from the beds of Torres Strait. North Australia."[32] The print also has a photo agency stamp on the back that reads "From Rapho-Guillemette Pictures."[33] In other words, this photograph and its textual explanation—which infers that it is a visual document of a diver on the floor of the sea at the Torres Strait—had a global life, and the potential to reach hundreds if not thousands of viewers. Rapho-Guillemette Pictures was the New York branch of a significant early photo agency started in 1933 in Paris by Charles Rado (1899–1970), along with founding members that included the surrealist photographer Brassaï.[34] That Hurley's work was handled by Rapho-Guillemette Pictures demonstrates how central he was to the evolution of modern media culture and how easily accessible his work was to the avant-garde of Europe, particularly the surrealists. The agency specialized in humanist subjects and was involved in the aptly named *Family of Man* exhibition staged in 1955 by Edward Steichen for the Museum of Modern Art in New York. To show the breadth and variety of human experience was the aim of many early photojournalist agencies. Their objective was to take documentary photographs of importance to the public, and the public, in return, expected authenticity. Yet many news photographs were staged.

Rapho-Guillemette Pictures handled the international distribution of the photograph in figure 10.2, but the same image was also published widely in Australia. It appeared in the Melbourne *Argus* in 1933, without fish, and with a caption that reads, "The pearl diver—an under-water photograph of a diver collecting shell in Torres Strait."[35] Yet when it appeared in 1950 in Hurley's book *Queensland, A Camera Study*, it was published, as in figure 10.2, with fish.[36] The same image was reproduced again without fish in an article by Hurley for *Walkabout* magazine titled "The Pearl Divers of Torres Strait" in which the caption said that the photograph was "taken under 60 feet of water."[37] The image and caption therefore ask the reader to picture Frank Hurley sixty feet beneath the surface of the Torres Strait, very close and at eye level with a pearl diver. Sixty feet beneath the surface of the Torres Strait, though, underwater currents were so strong that divers were barely able to bend to collect pearl shells.[38]

Dennis Doordan asks an interesting question when he inquires whether the artificial environment of an aquarium can evoke the same wonder and astonishment as wilderness.[39] The popularity of Hurley's artificial image of an underwater Torres Strait diver suggests that it can. But the image was an exciting spectacle regardless of whether the public accepted it as a document or suspected it was fiction. Did they suspend disbelief in order to immerse themselves in the image? Did they believe that Hurley was a practiced deep-sea diver, and an underwater photographer, who had photographed pearl divers at the bottom of the sea? Did they have a naïve response, or did they acknowledge the trickery and showmanship? In Hurley's case, it was to his advantage that viewers were inexperienced with reading underwater photographs, and lacked knowledge of the appearance of underwater ocean environments—it added to the mystery and intrigue of Hurley's work. As Dona Schwartz argues, there are certain conditions of viewing photographic images in which it can certainly be said that viewers are naïve because they "have not learned the cultural conventions that facilitate the process of interpretation."[40]

Hurley had integrated the pearl diver metaphor into his own public image: he was the lonely figure setting out to the mythical unknown and returning with treasure in the form of photographs and films. He told the public in 1926 that he had the technological capacity "for taking the deep-sea scenes depicting divers gleaning their harvest of pearl shell from the depths of the Coral Sea."[41] But photographing a pearl diver at sixty feet was an impossible dream for Hurley. Through combination printing and studio methods, though, he found a way for viewers to feel and sense the undersea world of pearl divers, and for them to see him as a heroic creator of new visual perspectives on the unseen and the unknown. About diving mobility and underwater filmmaking in the 1920s, Franziska Torma has recently observed that underwater photography was a way for explorers to immortalize themselves in the quest for frontiers of visibility.[42]

Frank Hurley is famous for staging photographs and combining negatives in creative ways to make composites "to create pictures that could convey the drama of the human story of exploration."[43] Using two or more negatives, he superimposed images to transform subjects from the common to the awesome, and from staged and artificial to convincingly natural. As a filmmaker, it came naturally to Hurley to shoot scenes from sets built in the studio and to entice viewers to imagine the setting as authentic. In the First World War, as Martyn Jolly shows, Hurley superimposed heraldic skies over the carnage of battlefields. But while Hurley believed it enhanced the impact of the war photograph, the underlying artificial nature of the imagery drew moral outrage from the establishment.[44] Yet in the culture of the copy, as Hillel Schwartz calls modernity, the history of documentary photography and filmmaking is a history of simulation.[45] In Hurley's day, Walter Benjamin theorized modern media culture as a new world of technologically produced "synthetic realities."[46] The beauty of techniques such as splicing, composites, cropping, and framing lay in the way they assisted Hurley to achieve the effect of realism, rather than mirror reality. It served his primary purpose, which was to tell exciting stories through pictures.

Composite printing was fairly standard practice in the circles Hurley moved in, including among explorers. For example, Hurley's American business associate, Lowell Thomas (1892–1981), who was a renowned explorer, journalist, photographer, and public lecturer, also composed effective, but fictitious, news photographs. He told the story of the time he combined two negatives to create a photograph "documenting" an interview with Theodore Roosevelt, an interview that never took place.[47] Hurley met Thomas in Palestine during the First World War. Early in 1921, his plan was for Thomas to deliver synchronized lectures for the American season of *Pearls and Savages*, but that plan fell through, and Hurley himself decided to travel with the film and speak in public. Thomas was a celebrity travelogue presenter and member of the Explorer's Club in New York.

The Explorers Club, a gentleman's club with a distinguished membership that included Theodore Roosevelt, Carl E. Akeley, and William Beebe, was formed in 1904. A recent history of the Explorers Club explains how its mission was "spurred by the challenges of reaching the unreachable and the scientific desire to pry from the earth its long-held secrets."[48] It thrived in the era that Richard Conniff calls "the great age of biological discovery [when] nations, museums, and universities all sent out collecting expeditions."[49] Its members explored the dark and dangerous corners of the globe, and returned with stories, artifacts, and images that gave people at home a reassuring sense of the economic, aesthetic, technological, and moral superiority of white civilization. Exploration could be a lonely life, but among the rewards was deep respect from a male community. From the perspective of Richard C. Wiese, a past president of the Explorers Club, the late nineteenth century and early twentieth century was the "heroic age of exploration."[50] The world was a stage on which to test the cornerstones of masculinity: courage, risk taking, originality, and solitariness.

Hurley was not a member of the Explorers Club, although the work he undertook, especially in the Antarctic with Shackleton—who was a member—and also in the remoter territories of Papua, made him eminently qualified. New York was a long way from Australia, making membership in the club impractical for Hurley. However, many of the club's members were well known to him, among them the underwater photographer and filmmaker John Ernest Williamson.[51] There were many opportunities for Hurley and Williamson to cross paths, including through clubs and agencies, especially in the United States, where both men delivered film screenings and lecture entertainments. They were both affected by ethical debates surrounding fakery in representations of nature. How do Frank Hurley and J. E. Williamson compare in terms of careers, especially in relation to the main issues of their day pertaining to the reception of photography and filmmaking by explorers whose audiences were mainstream, avant-garde, and scientific?

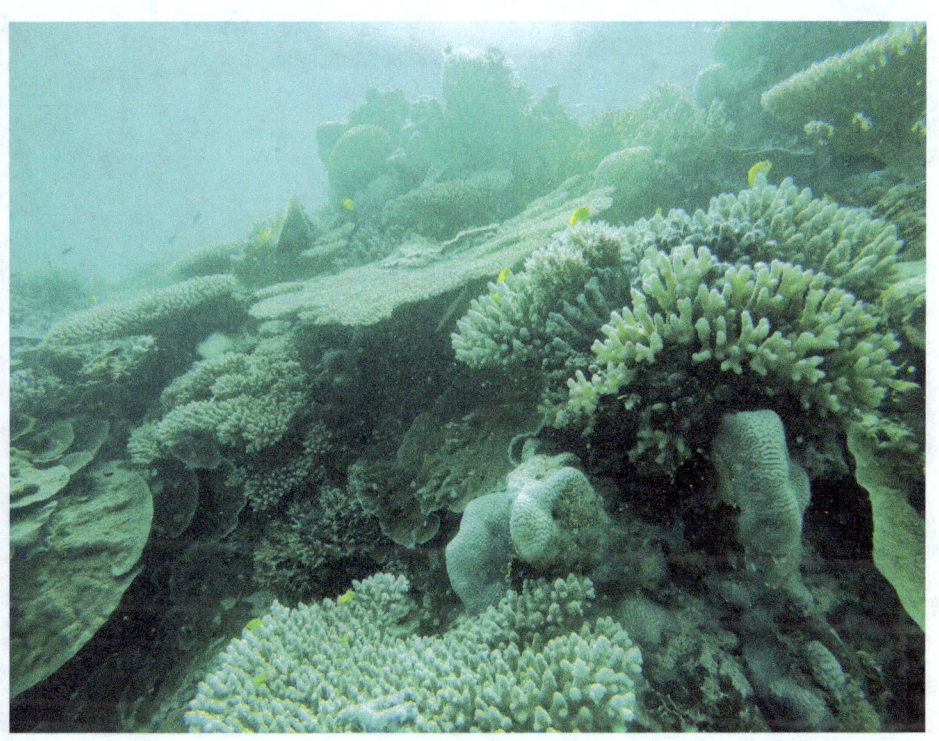

Plate 1. The Great Barrier Reef, 3D Reefs Program, 2017. Photo by Will Figueira.

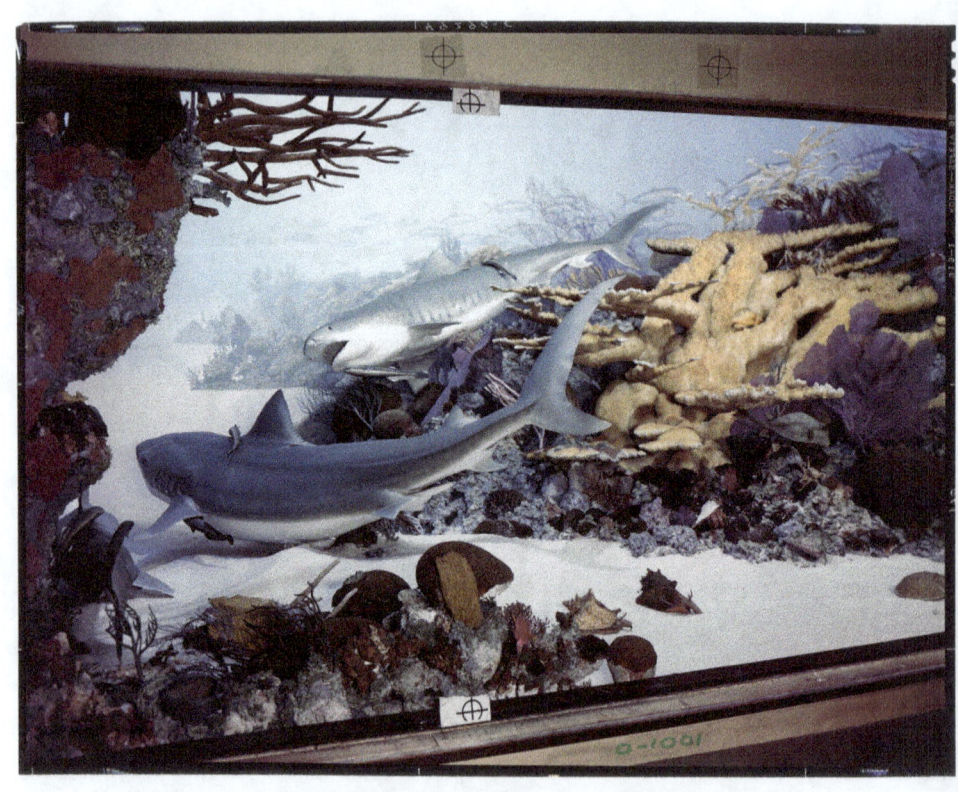

Plate 2. Bahama Islands diorama (left section), Field Museum, Chicago, 1929.
Courtesy of the Field Museum, z15т.

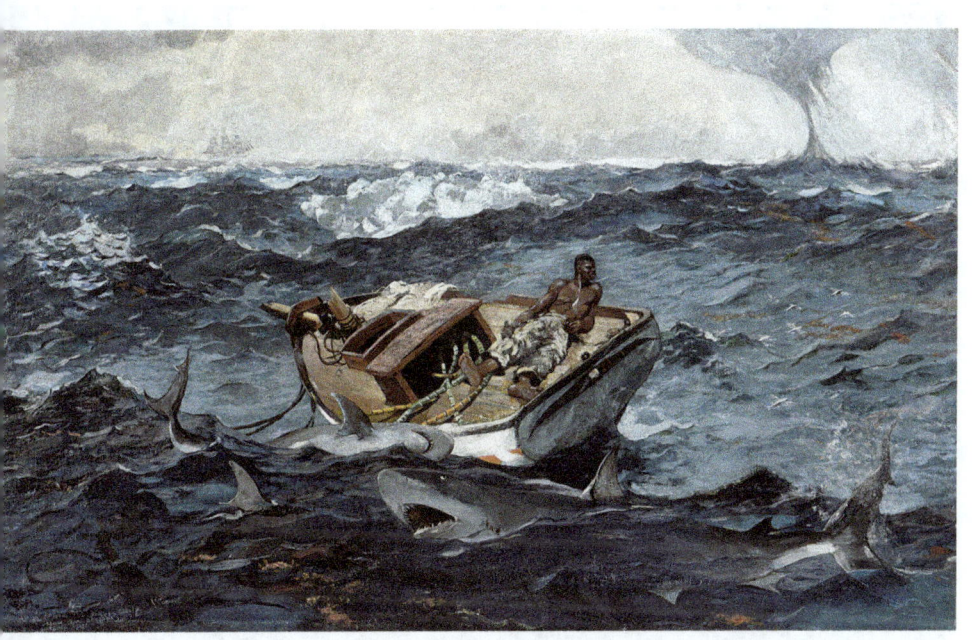

Plate 3. Winslow Homer, *The Gulf Stream*, 1899. Oil on canvas, 71.5 × 124.5 cm. The Metropolitan Museum of Art, Catharine Lorillard Wolfe Collection, Wolfe Fund, 1906 (06.1234).

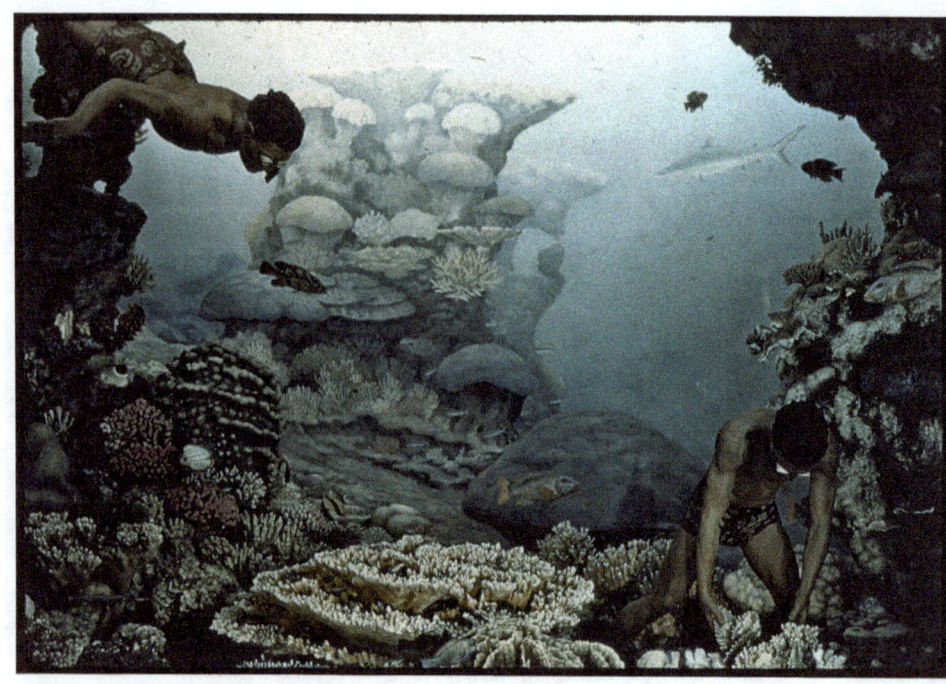

Plate 4. "Pearl Divers in Coral Reef, Tongareva, French Polynesia"
diorama, Milstein Hall of Ocean Life, 1937–1941. Image K1996,
American Museum of Natural History Library.

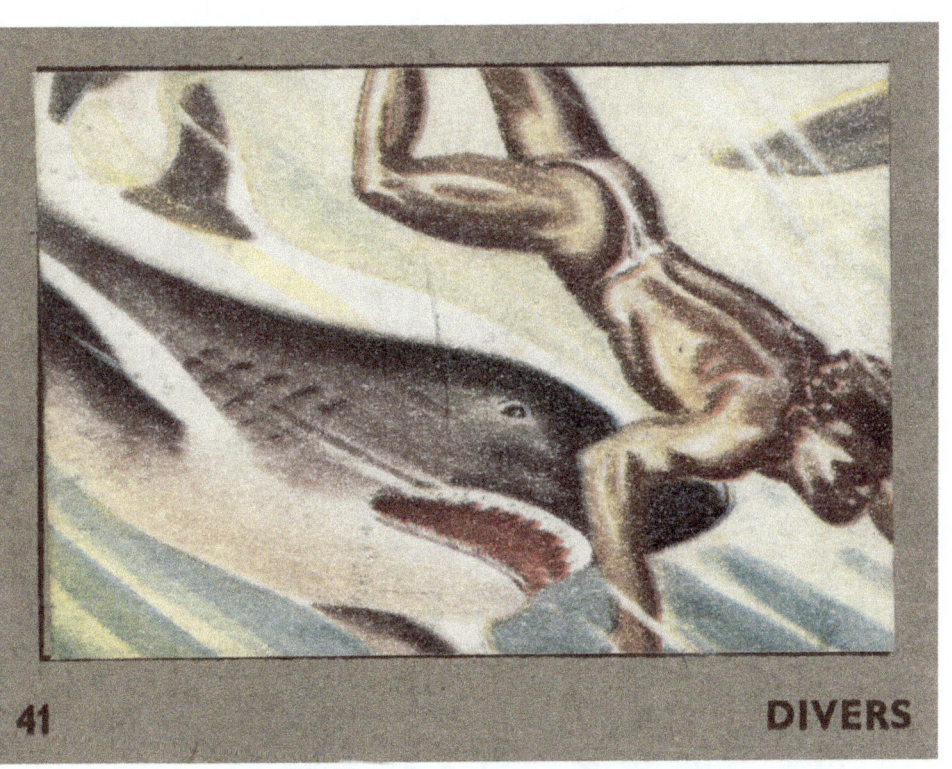

41 DIVERS

Plate 5. "Divers," *Marvels of the Great Barrier Reef* (No. 41), Sanitarium Health Food Company, 1948. Courtesy of Sanitarium Health and Wellbeing.

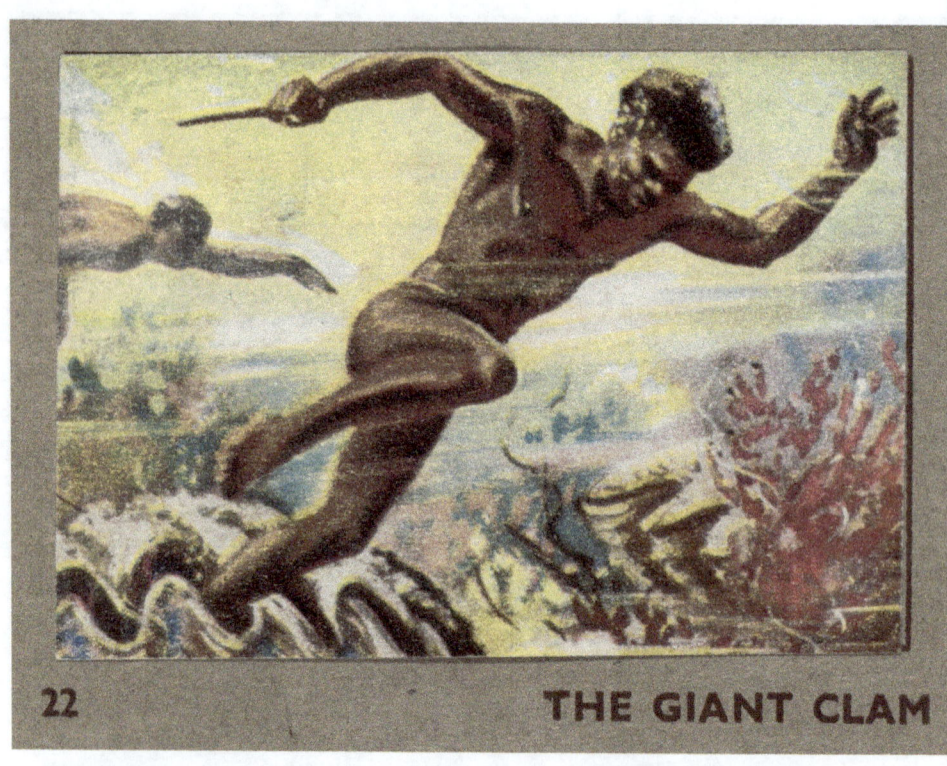

22   **THE GIANT CLAM**

Plate 6. "The Giant Clam," *Marvels of the Great Barrier Reef* (No. 22). Sanitarium
Health Food Company, 1948. Courtesy of Sanitarium Health and Wellbeing.

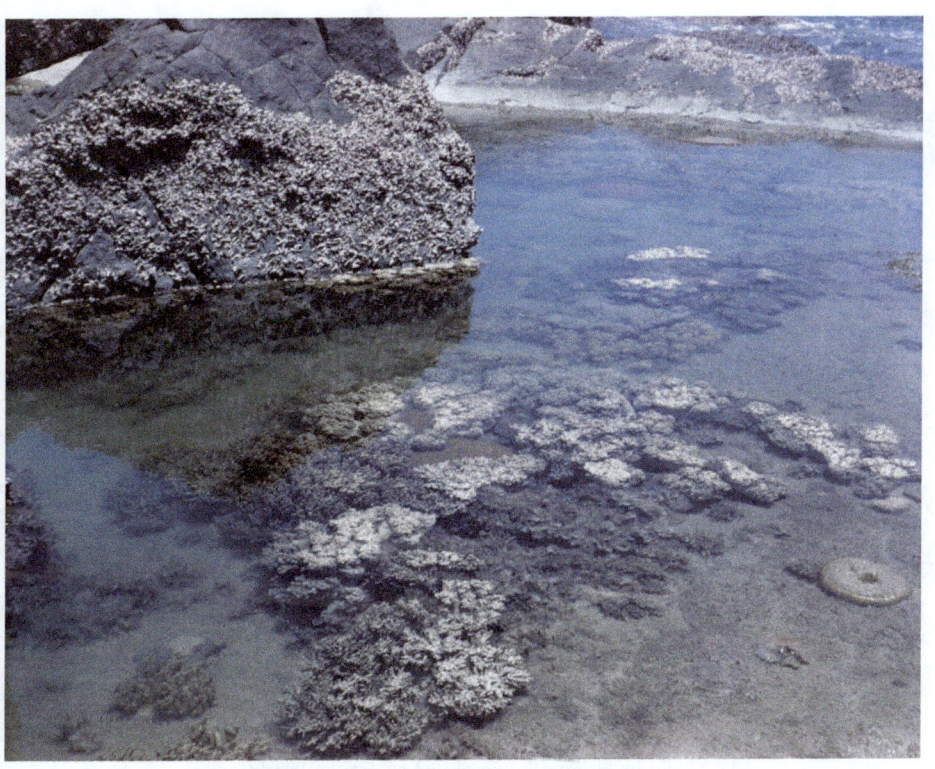

Plate 7. Frank Hurley, "Coral Reef," Great Barrier Reef, 1921.
National Library of Australia, NLA CDC-10487289.

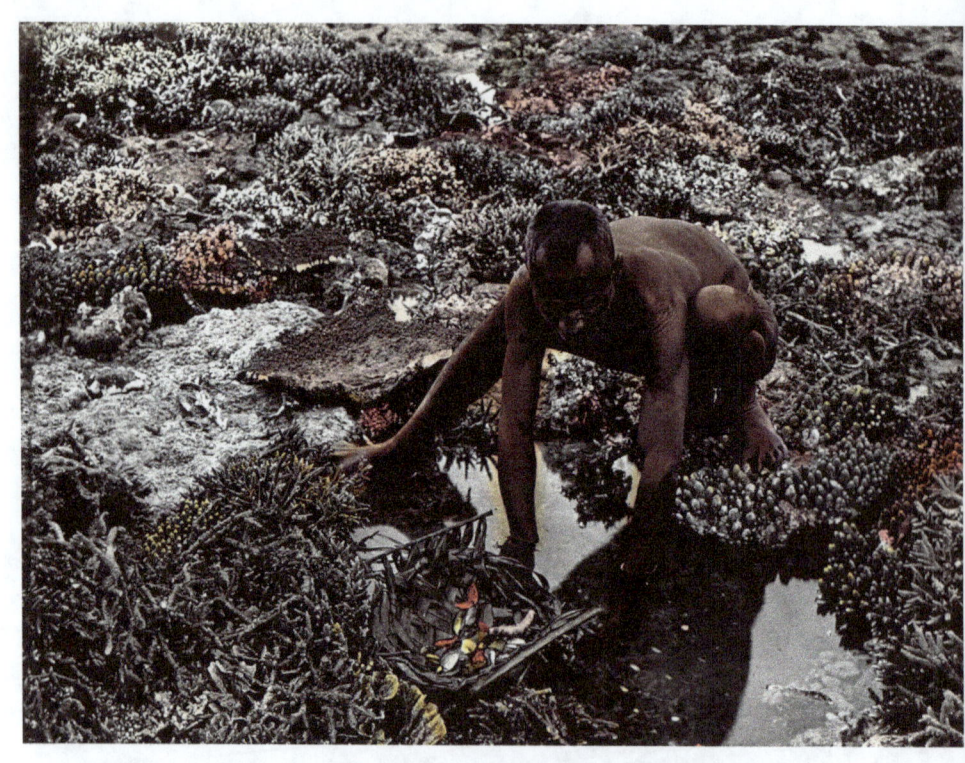

Plate 8. Frank Hurley, the coral reef off Dauko Island, 1922.
National Library of Australia, PIC FH/839.

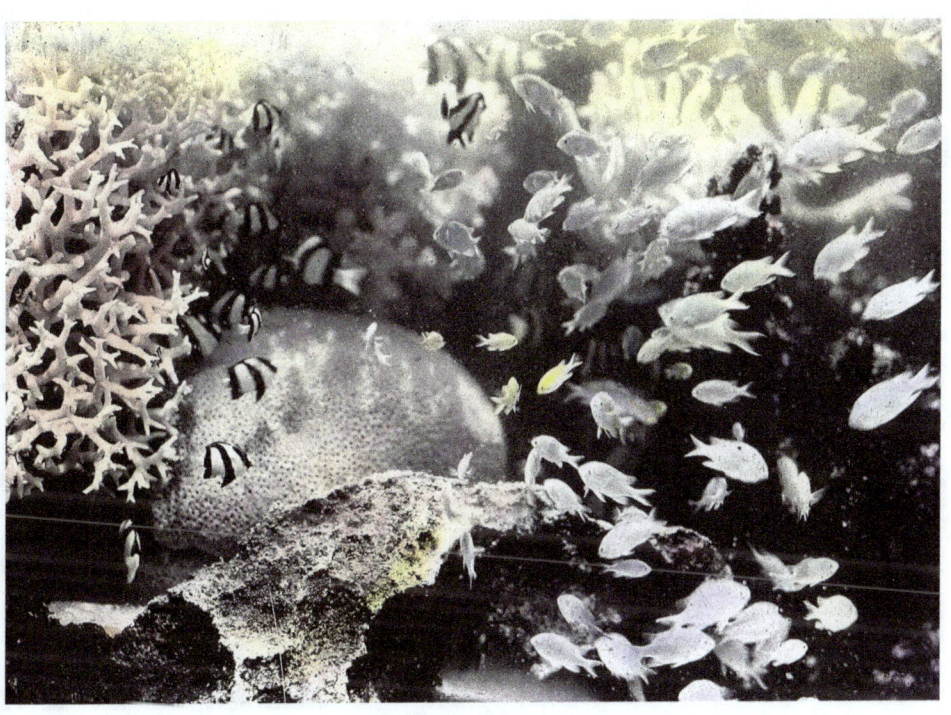

Plate 9. Frank Hurley, fish underwater, 1922. Colored lantern slide.
Courtesy of Australian Museum Archives, Frank Hurley Photographs
AMS320/V3242.

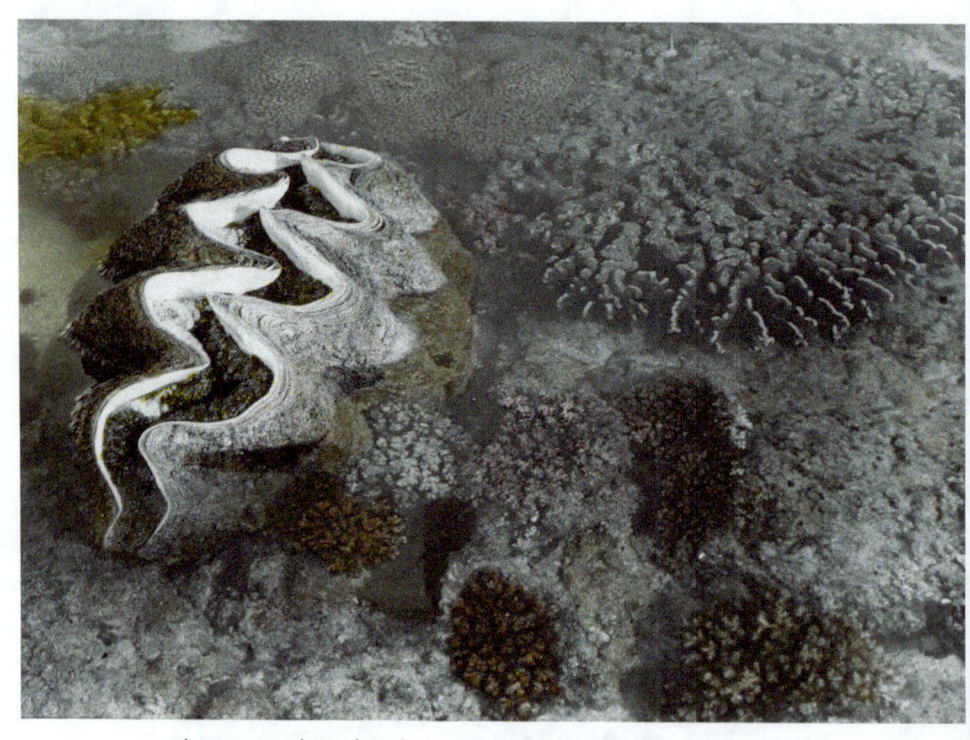

Plate 10. Frank Hurley, clam and corals, Great Barrier Reef, ca. 1921.
National Library of Australia, PPIC FH/8990.

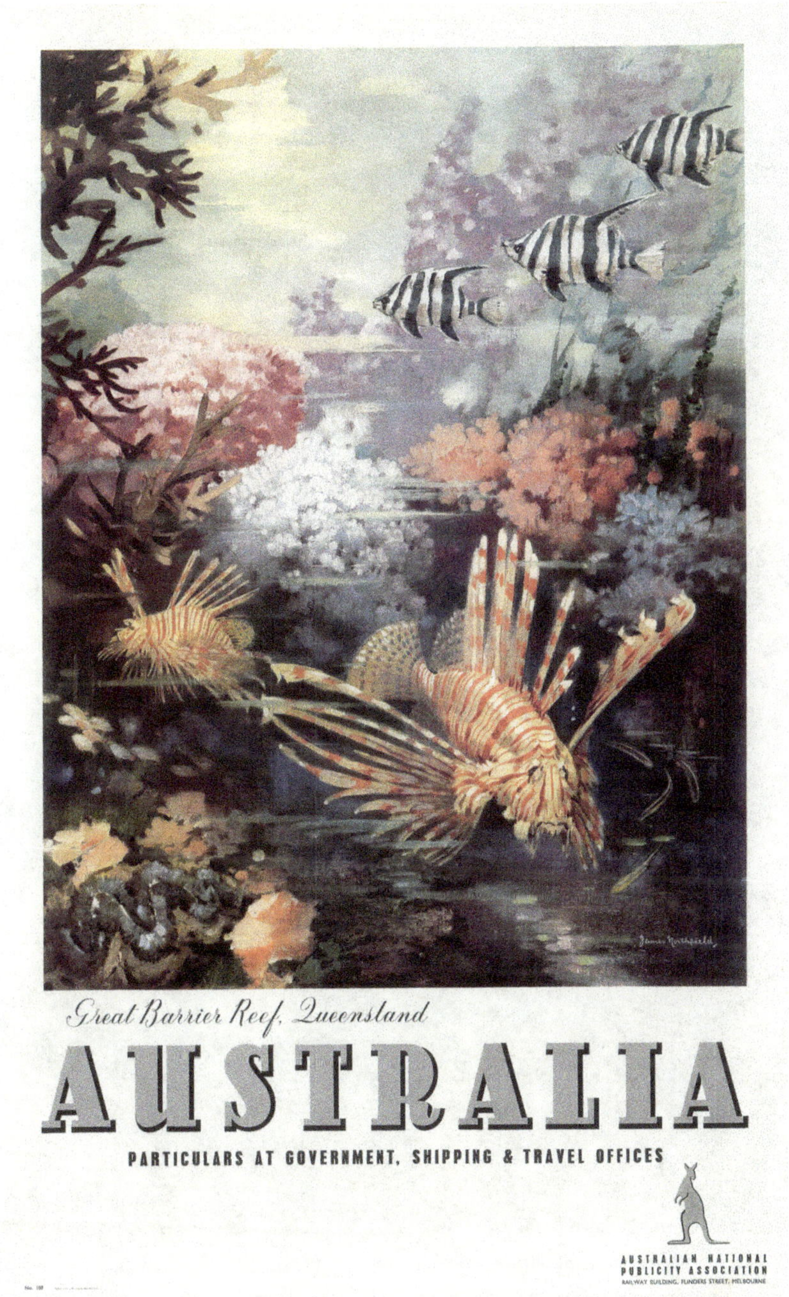

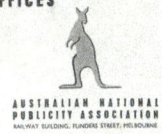

Plate 11. James Northfield, Great Barrier Reef Queensland, poster,
101 × 63 cm, ca. 1935. National Library of Australia, PIC LOC DRAWER 216,
© James Northfield Heritage Trust.

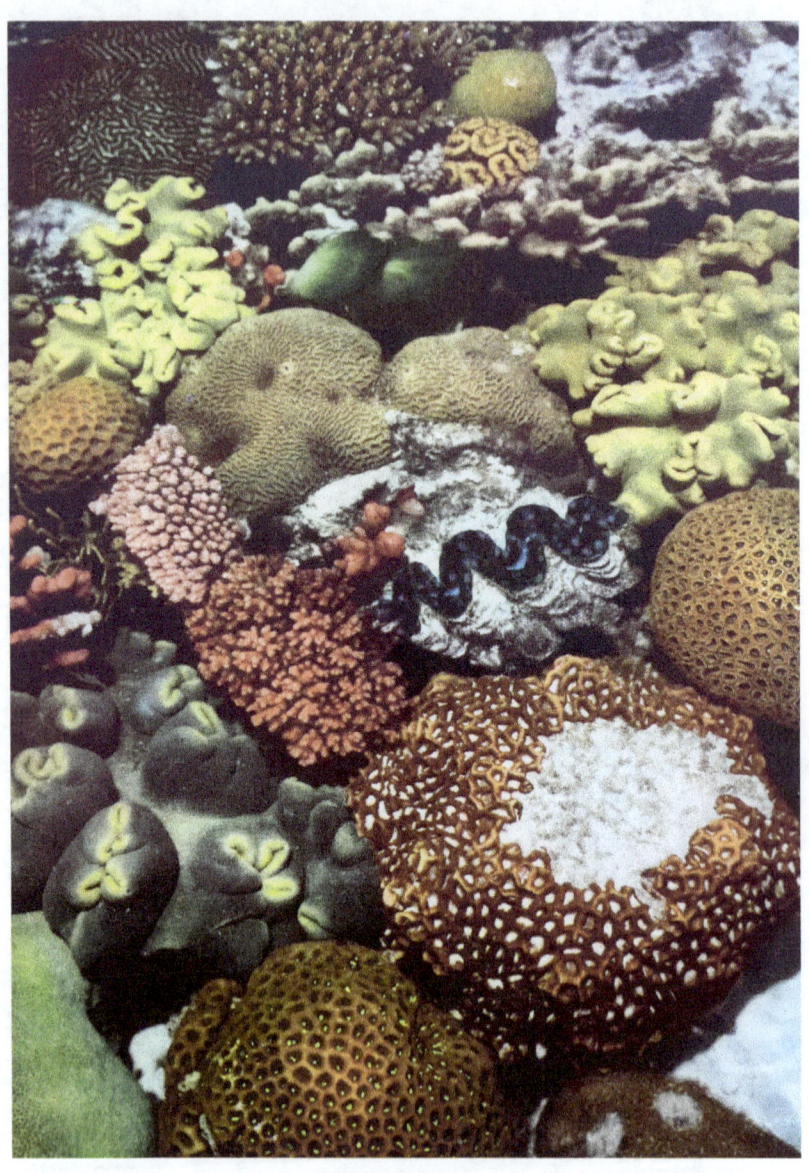

Plate 12. Frank Hurley, "A harmonious group of soft and stony corals seen through the transparent water," ca. 1950. From Frank Hurley, *Australia: A Camera Study* (1955), 94.

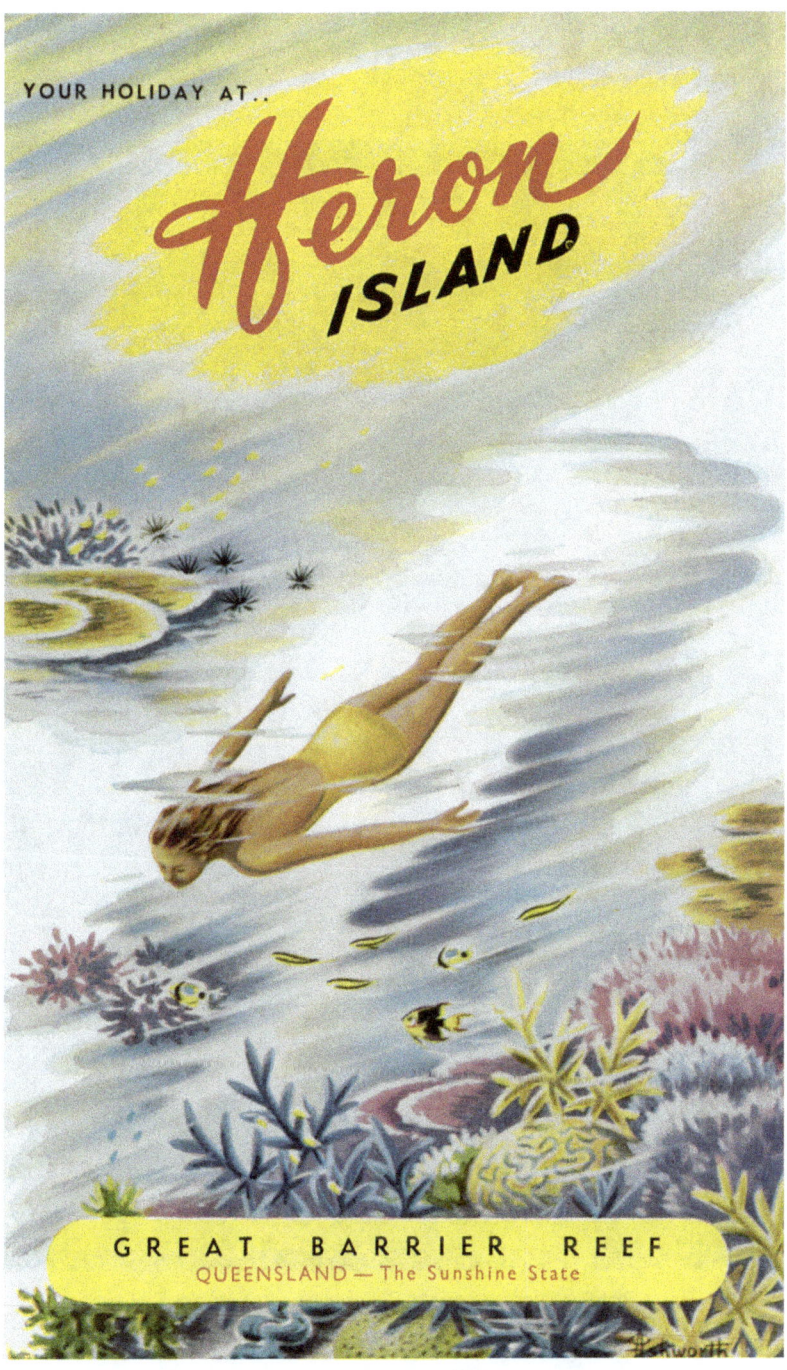

Plate 13. Heron Island holiday brochure, Great Barrier Reef, Queensland tourism, 1956. National Library of Australia, ephemera collection 4853702.

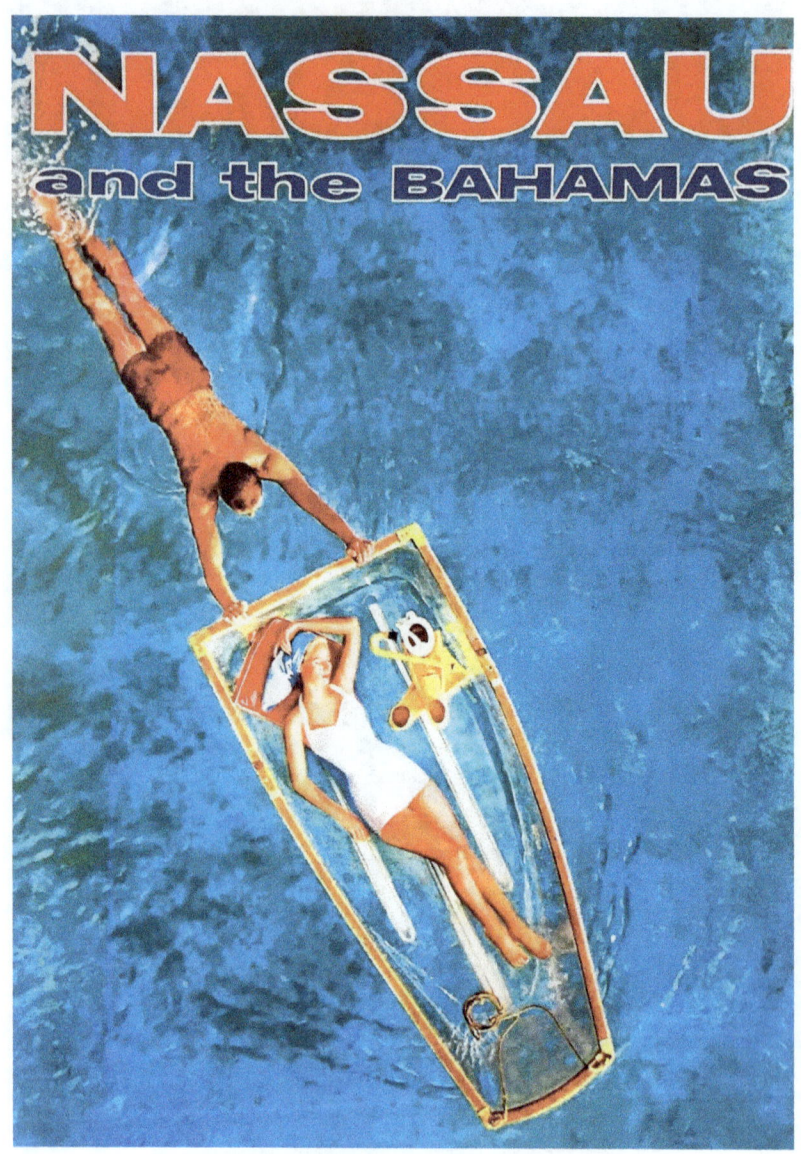

Plate 14. *Nassau and the Bahamas*, poster, ca. 1959.

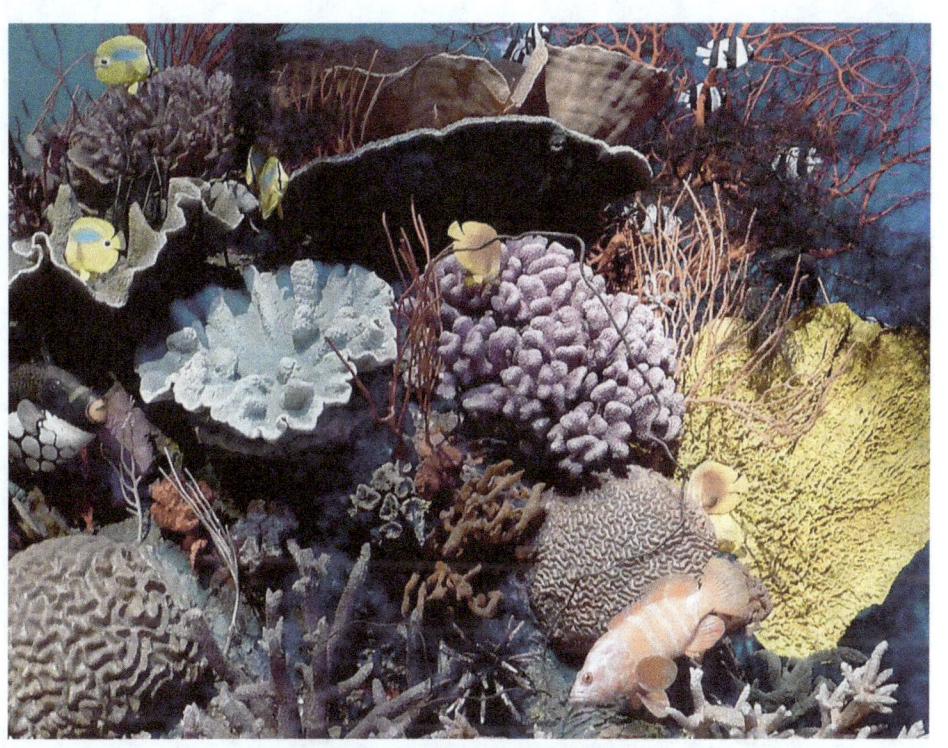

Plate 15. Peter Peryer, *Coral Reef*, 2013.
Digital photographic print, 490 × 660mm.

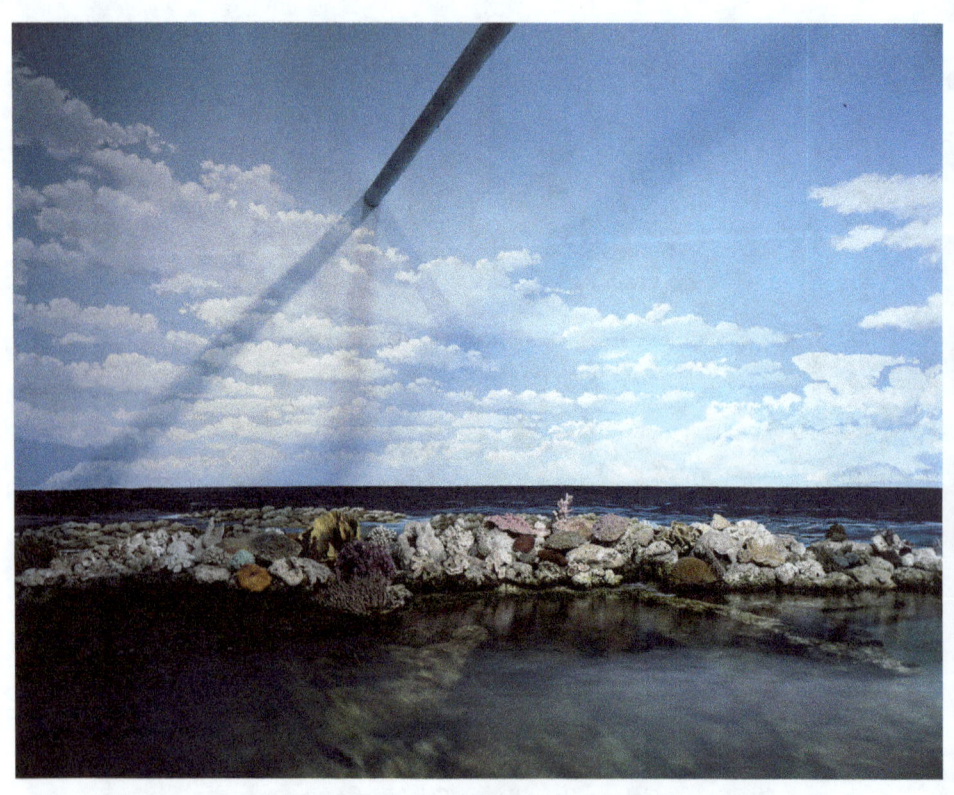

Plate 16. Frederico Câmara, *SEA LIFE Sydney Aquarium, Sydney, Australia*, 2014. Photograph, 150 cm × 120 cm. Courtesy the artist.

PART IV

Hurley and Williamson

CHAPTER 11

Explorers and Modern Media

According to Marshall Berman in 1982, the contradiction of being modern has always been "to find ourselves in an environment that promises us adventure, power, joy, growth, transformation of ourselves and the world—and at the same time, that threatens to destroy everything we have, everything we know, everything we are."[1] At the Explorers Club in New York in the early twentieth century, where distinguished men such as Theodore Roosevelt and Carl Akeley met and exchanged stories, ideas, and camaraderie, exploration was a heroic undertaking arising from the power of authentic accounts about the natural world, and from risks and dangers faced in the pursuit of truth and knowledge. But at the same time, there was something that threatened to destroy the very values that made explorer culture, and its associations with museums and science, great. It was a threat perceived in the progress of modernity itself, especially the growth of popular culture, the untrustworthy realm of mass media, including the superficial spectacle of visual images—especially moving pictures—and the socially dangerous realm of showmen who engaged with audiences through bluff, deception, and exaggeration. For many members of the elite Explorer's Club, modernity's dark side was the expansion of popular media culture. At risk through the mixing up of mass culture with the seriousness of science and exploration were ideals of masculinity, knowledge, and morality.

In the early twentieth century, Frank Hurley and J. E. Williamson were in a curious situation: as explorers, their task was to expand science through knowledge, but as entertainers their ambition was to excite the public imagination with films and startling images. If anything, the social relevance of explorer culture had started to fade by 1920, but Hurley and Williamson never relinquished their claim on that unequivocally heroic and male domain. Even so, as producers of the new reproducible media culture, as participants in and shapers of modern visual practices and entertainment, their work was also implicated in what was seen as the devaluing of "authenticity," "originality," and "first-hand experience."[2]

Consequently, early twentieth-century modernity was both a difficult and an exhilarating landscape for Hurley and Williamson to operate in. Two separate events, one for Hurley and another for Williamson, will explain why this was so, and help illuminate the often-bewildering contradictions and conflicts they faced. Both incidents involved Carl Akeley, the eccentric but talented and celebrated curator, taxidermist, and filmmaker who helped to make the Field Museum in Chicago and the American Museum in New York among the Western world's greatest institutions of natural science.

In sum, the events were sparked by increasing animosity toward filmmakers who were seen to tip the scales of acceptability in the area of wildlife documentaries by ignoring conventional boundaries between education and entertainment, and by seeking to exhibit works of spectacle in science museums. Of the two, it was Hurley more than Williamson who was seen to cross that line. However, both men were connected either directly (for Williamson) or indirectly (for Hurley) with "nature faking." Accusations of "nature faking" were reserved for the many early wildlife films that professed to be shot in authentic ecological context and that claimed to show the natural habits of wild animals but were instead filmed in zoos, aquariums, cages, and other sites of captivity.

While those who faked nature may have been viewed as progressive by those who saw the inherent fiction-making properties of film as an asset, they appeared regressive to members of the Explorer's Club, where reality denoted something honest and true. "Nature fakers" represented all that was degenerate about the modern world, particularly to club members such as Theodore Roosevelt, who stated, "As real outdoor naturalists, real observers of nature grew up, men who went into the wilderness to find out the truth, they naturally felt a half-indignant and half-amused contempt for the men who invented preposterous fiction about wild animals, and for the credulous stay-at-home people who accepted such fiction as fact."[3]

In the gendered history of explorers and outdoor naturalists, of men in the 1920s who did as Theodore Roosevelt said (and believed was vital), who "went into the wilderness to find out the truth," Frank Hurley and J. E. Williamson walked the fine line between exemplary masculinity and compromised masculinity. At a time when it was thought that only men of lesser moral fiber produced wildlife films and travelogues for popular entertainment by fabricating scenes of animals and wilderness, or by passing off their work, designed for entertainment, as "scientific," there was sometimes room to doubt the quality and honesty of both Hurley and Williamson.

Hurley and Williamson could be said to have exhibited the ultimate signs of manliness—commanding with machines, proficient with engineering, gifted at exploring new frontiers.[4] Both born in the nineteenth century, they were raised to enjoy and respect what Hurley once called "rude adventure and romance."[5] Their early lives were governed by stories, myths, and illustrations by, and about, imperial explorers such as James Cook, literary visionaries such as Jules Verne, and the many authors of sea adventure fiction who Margaret Cohen discusses in *The Novel and the Sea* (2010), characters who "battle life-threatening storms, reefs, deadly calms, scurvy, shipwreck, barren coasts, sharks, whales, mutinies, warring navies, natives, cannibals, and pirates—in short they have *adventures*, as many such novels emphasized with the wording of their titles."[6]

As rude adventurers and romantics, Hurley and Williamson went exploring for the extraordinary and the uncanny: lost tribes, lost worlds, the unknown deep, and the planet's hidden secrets. But in order to make their achievements known, they utilized modern communication systems, produced films and photographs, presented lecture entertainments in large public venues such as the Smithsonian Institute and the American Museum of Natural History, and toured the world's biggest cities in the United States, Britain, and Canada. To satisfy their audiences they spent their lives chasing images both startling and sublime, and they consolidated their positions as inventors and explorers by writing for the popular press. Their self-promotion was tireless, and both were great storytellers, which confirms Edward Said's argument that stories "are at the heart of what explorers and novelists say about strange regions of the world."[7]

But Hurley and Williamson practiced at a time of increasing concern about culture being feminized by the tricks and deceptions of mass media, and by a burgeoning modern industry catering to what seemed like passive and undiscerning spectators seeking entertainment rather than truth, as alluded to in the quote by Roosevelt above. Driving the social changes brought about by consumer culture was a new class of professional that authorized and encouraged the reproduction of images, information, advertising, and publicity, and the growth of illustrated magazines and newspapers. To an older and more traditional generation, the new class of professional was responsible for weakening culture and for making experiences of nature secondhand, manipulated, and inauthentic. Mike Featherstone points out how the expanding culture of the copy alienated citizens who valued the ideals surrounding nineteenth-century models of heroes: "extraordinary deeds, virtuosity, courage, endurance and the capacity to attain distinction."[8]

The underwater in the early twentieth century was a place to achieve extraordinary deeds and display courage and distinction. Which is why the globalizing media world of the first three decades of the twentieth century—press agencies, photo agencies, and lecture-circuit agencies—capitalized on the stories and pictures of the underwater frontier that Hurley and Williamson obtained through exploration. Hurley collaborated with Lowell Thomas Travelogues to distribute his films overseas, but in the United States his agents were also Lee Keedick and James B. Pond. Keedick was also J. E. Williamson's agent for North American lecture tours. Keedick advertised his business as being "Manager of the World's Most Celebrated Lecturers."[9] He promoted Williamson as "Mr. J. E. Williamson: Distinguished explorer and originator of undersea photography and motion Pictures."[10] Originator of undersea photography? The claim was exaggerated, but hyperbole was essential to the competitive world of celebrity explorers. The titles of Williamson's lectures were also carefully chosen to seize attention: "Into the New World under the Sea" and "Beauty and Tragedy under the Sea."[11]

Hurley was a client of Topical Press Agency, Rapho-Guillemette Pictures, and Universal, while Williamson was a client of Wide World Photos, an association, as we have seen, that brought Williamson's underwater imagery to the attention of the French avant-garde. Their images were published in the same mass-produced magazines, including the *Illustrated London News*, and a popular set of volumes comprising lively essays and striking photographs of the natural and man-made worlds titled *Our Won-*

derful World. The subtitle of *Our Wonderful World* signposts the aim of the publication: to show the public the "marvels of nature and the triumphs of man."[12] The publication focused on stories about man in the realm of nature, on the technological marvels of man in nature, and on comparisons between the marvelous features of the man-made landscape and the natural landscape. In other words, the publication's real interest was in the triumph of man over nature rather than nature as inherently and independently fascinating. But the idea of triumphing over nature that underpinned the consumerist philosophy of *Our Wonderful World* was also the common ground that Hurley and Williamson shared: it was a sign of their being modern.

Reading *Our Wonderful World* and *Popular Mechanics*, two magazines of popular science, is a way to gain insight into the place of Frank Hurley and J. E. Williamson in modern pictorial culture, and to gain a perspective on why their photographic representation of coral reefs and underwater environments were seen as modern. *Our Wonderful World*, more than other contemporaneous publications, addressed the kinds of ocean topics that were important to both Hurley and Williamson, including "the wonders of the coral island," "the wonderland beneath the sea," and "the trials and triumphs of deep sea diving." There is an emphasis on articles featuring underwater exploration, and often Williamson's photographs were featured showing corals, small fish, and sharks outside the window of the photosphere and underwater in the Bahamas. His images contributed an exciting, futuristic, adventure-oriented tenor to a journal that saw itself at the cutting edge of science for all, in a publication that combined articles by professional scientists with submissions by amateurs.[13]

Modern magazines and journals of popular science writing offered a natural home for Hurley and Williamson, who were able to publish nonspecialists' viewpoints alongside those of science professionals. *Our Wonderful World*, *Illustrated London News*, and *Popular Mechanics* were regular venues for their photographs, although often the photographs were loaned by photo agencies and gave no attribution to the photographer. This was the case with Frank Hurley's photograph of a giant clam (see plate 10), a photograph taken during the 1921 expedition to the Torres Strait and Papua. In *Popular Mechanics* in 1924, the clam photograph was attributed to Hurley, and the photo agency was given as "International."[14] But in *Our Wonderful World* in 1933, Hurley's name was not mentioned, only the agency "Topical Press."[15] With photo agencies in charge of the distribution of images, photographs often had a life of their own separate from original contexts and makers.

It was around 1910 that a new type of documentary film emerged—the "documentarist-as-explorer."[16] There was scope within this emerging genre of filmmaking for emotional tension and excitement to sit alongside truth to nature. The "documentarist-as-explorer" was the cultural space that Hurley and Williamson came to occupy, a genre that opened up possibilities for embellishing the truth. But it was this genre that caused controversy through associations with deceptions and fakery. Nature faking was a heated public debate in the first decades of the twentieth century. Faking nature put at risk the integrity of natural history as a study and representation of the facts of nature, and exploration as a journey into wilderness to study nature honestly. The new class of professional, that of agents and promoters who handled images, information, advertising, and publicity, who responded to modern life and public demand for entertainment through visual culture was implicated in the problem. To fake nature in films and photographs for mass entertainment and profit was to debase higher moral principles such as courage, and it was, as Gregg Mitman explains, a threat to men of Roosevelt's and Akeley's class because the threat it posed was "the physical and moral effeminacy of modern times."[17]

In the cultural sphere of the early twentieth century, cinematic tricks only served to amplify concerns about humanity's increasing inability to distinguish illusions from reality. Cinema, it was said, made prisoners of people in a shadowland, like the prisoners in Plato's parable of the cave who could not distinguish real objects from their shadows or doubles. So, while cinema in the mid-1910s was the dominant form of popular visual entertainment, upholders of "high" culture condemned it as a "mechanical and vulgar carnival attraction."[18] Cinema's popularity exacerbated a fear that machines were diminishing the triumphs of man by creating superficial doubles to replace real life.

With the rise of cinematic representations of nature in documentary films came unresolvable questions about how to recognize borders between reality and illusion, as well as fiction and science. After the turn of the twentieth century, documentary films about nature were, according to Erik Barnouw, "infected with increasing fakery," the result, he suggests, of growing competition among filmmakers.[19] Consequently, not all wildlife documentaries were shot in situ. Barnouw cites the case of a Danish "safari" documentary in which jungle footage was "intercut with close shots" of African animals in the Copenhagen zoo.[20] But people who were science purists believed "authentic" wildlife films should be made by traveling to the home of the animal, whatever the conditions, to capture film footage, photographs, and drawings of the animal's ecological context.

There are many photographs, for example, of Carl Akeley and his helpers on location in Africa preparing drawings of the environment for future reference when designing dioramas. Akeley had implicit faith in the unity of science, fact, and documentary. It was that faith that put him at odds with popular, mass media culture and the spectacle of showmanship, entertainment, and fiction. Akeley was a filmmaker, and his commitment to scientific truth in filmmaking was so deep that he designed a special portable cine camera for natural history filmmakers that was unique because it could capture panoramic views.[21] In other words, it could capture a view that revealed the seamless actualities of ecological context. Roosevelt and Akeley, who opposed "nature faking," argued that the quest for truth and fact was a sign of social maturity, whereas those who created fictions about nature were immature and socially irrelevant.

Looking back on that time in history, Donna Haraway argues that the fear of nature faking was a fear of the "failure of order" of class, race, and sex, a stable social order based in ideologies that determined what was real and true, and a social order that had been in place for centuries. Never before had that social order been so undermined as it was by modern society's increasing inclination toward simulation and deception, by the social challenges presented by the emancipation of women, and by increasing agency sought by nonwhite peoples. In light of claims by museum scientists such as Akeley that the growing spectacle of photography and filmmaking was a threat to the authentic experience of nature as a living, breathing phenomenon, Haraway notes that taxidermy—the practice of which Akeley was considered the master—anticipated the "virtual touch of the camera."[22] Taxidermy was also responsible for transforming nature into an image for consumption.

The striking lifelike dioramas and displays for which Carl Akeley was admired were the outcome of years spent inventing cunning contrivances and sophisticated illusions. The dioramas he built, including "Fighting African Elephants" at the Field Museum, are elaborate spectacles that entertain as much as educate. In the pursuit of realistic dioramas, Akeley even invented a cement gun for the purpose of creating imitation rocks and mannequins.[23] Reflecting on Akeley's celebrated taxidermied animal exhibits, especially dioramas of African elephants, Bernd Behr observes, "If Akeley's realism is celebrated for its exacting faithfulness to nature, then the dramatic staging of his dioramas show that realism is also a highly scripted scenario that says more about the culture that produced it than the "natural" subject it depicts: the skin is grafted onto an intended image, the index yields to an ideal."[24] Even in an age of enlightened knowledge, Akeley's dioramas were intimately related to the emergence of modern entertainment industries, and in this regard helped shape and change modern society by inventing nature as a visual spectacle.[25]

In 1936, when J. E. Williamson assessed the contradictions, inconsistencies, and double standards of museum scientists, he reached a similar conclusion as Haraway: scientists like Akeley were also entertainers. Yet despite the soundness of this reasoning, nature faking was not an association that Williamson wanted for himself. It demeaned his achievements, and threatened to make his enterprise seem significantly less serious, and less masculine than was ideal. It made him seem less of an explorer: "I hope the average man does not imagine that all my job consists of is to sit in my armored shell and turn a camera crank. Apart from having a sound knowledge of photography, the marine explorer must be versed in seamanship, navigation, geology, natural history, and other sciences. He must know where certain fish are to be found and where specimens of coral and marine vegetation may be located."[26]

Williamson didn't want to be known as a mere technician or an entertainer. However, the controversy surrounding nature faking was something he did get caught up in. In 1916, when Universal Pictures released Stuart Paton's 20,000 Leagues under the Sea, the problem of nature faking was at its peak. The topic of the "nature faker" was especially prominent in the press following the First World War, when camouflage, a technique of deception and fakery, was formally integrated into military operations. Creating perceptual confusions between the natural and the artificial became a modern war weapon, as when observation posts were disguised as fake trees.[27] Yet the new camouflage warfare represented a destabilizing force in society. Many prominent people, including Theodore Roosevelt, believed that in war, society, and nature, camouflage and hiding were signs of weakness.[28] But in the world of art and design, magic and illusion, business and entertainment, the strategy of concealing and revealing for gain and dramatic effect was more or less naturalized. It was, after all, through the deployment of artists for warfare in 1914 that the "science" of military camouflage came into being.

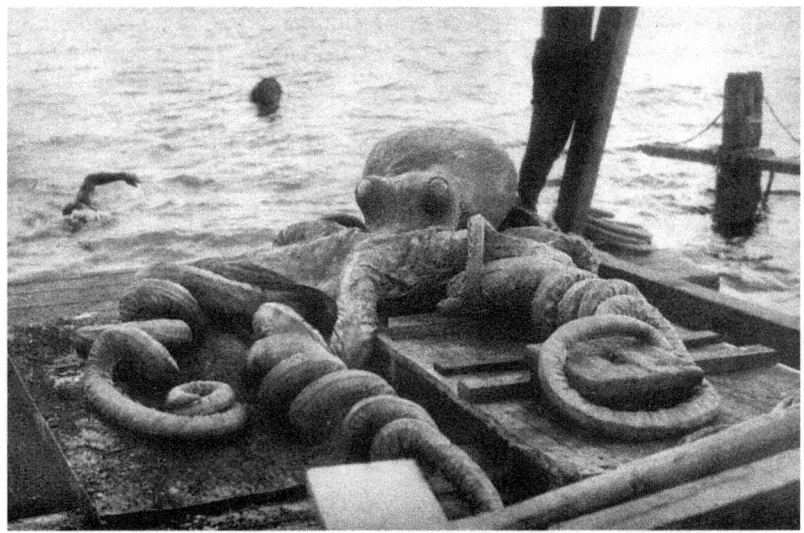

Fig. 11.1. J. E. Williamson, "The octopus at rest." From J. E. Williamson,
Twenty Years under the Sea, 1936, 17.

In 1916, for Paton's *20,000 Leagues under the Sea*, J. E. Williamson built
a gigantic mechanical octopus out of canvas and rubber tubing, a fake octopus that reached its monstrous size when inflated with air. The mechanical beast was controlled by a man sitting inside the creature's head.[29] It
was an ingenious invention, and what's more, it was strangely convincing.
When the movie was screened, audiences did in fact believe the animal
was real, and in the scene in which the creature seizes a "native diver" in
its giant tentacles, the sophisticated animatronics that created the apparition were put into action. With a demonic look in its eye, and with pliable
limbs, the octopus body appeared much less machine-like and more like
a living, breathing, muscular creature. Even out of the water, at rest on
the barge the *Jules Verne*, Williamson's robotic cephalopod looked lifelike
(fig. 11.1). Special effects were one of Williamson's true passions. He was
enthralled by *King Kong* (1933), especially because the gorilla automaton
"seemed to live and breathe and even to experience human emotions."[30]
He wanted to achieve something similar with the octopus. This is how
Williamson described the octopus scene: "The giant cuttle-fish glided
with sinuous motion from its lair. Loathsome, uncanny, monstrous, a very
demon of the deep, the octopus was a thing to inspire terror in the stoutest
heart. The native saw it. He turned—struck out for the surface."[31]

The scene is celebrated in the history of American silent cinema, and it is said that Williamson modeled the automaton on an octopus in the Brooklyn Museum to give it greater realism.[32] Audiences believed the octopus was real, but Williamson kept the fakery secret until 1936, when his autobiography was released. Before 1936, the public, including the Australian public, was told the monster was genuine.[33] Williamson did not want to disappoint audiences, and felt awkward that the deception had gone on for so many years, especially since he had maintained there was "no question of fake or deception."[34] In Australia in 1936, once the fakery was revealed, one reporter commented that looking back at the experience of watching Paton's film in 1916, the giant octopus was "so lifelike in its movements that [it was no wonder] the 'fake'-wise film critics of 1916 declared that it could not possibly be a 'fake.'"[35]

Instead of simply delighting his audiences, Williamson had deluded them, and it was that thought, that memory, that haunted him from 1916 to 1936, when the autobiography, along with a confession, was published. Perhaps, though, he overplayed the extent of the problem. Viewers may not have been fully able to explain the octopus, but it is difficult to imagine that every audience in every country where the film was screened failed to recognize that the octopus monster was mechanical, and then, having accepted it, did not enjoy the spectacle. The film, after all, was an adaptation of a work of fiction that belonged loosely to the "science fiction" genre, and was intended to appeal to the imagination as well as reason.

It troubled Williamson's conscience that he had misled the science community as well as the public about the octopus. But on reflection, he also questioned why he should feel guilty when it was doubtful there was much difference between himself and any museum scientist. Weren't they also showmen and entertainers, especially the taxidermists who built mannequins to look as lifelike as possible? Wasn't their work fakery too? Well might it be asked of habitat dioramas of the type Carl Akeley created: what is the difference between fiction and truth in these animal displays? Indeed, Williamson was probably thinking of Carl Akeley when he wrote, "Where is the dividing-line where the faithful recreation of any animal produced for any desired effect should be considered the act of nature faking? And where the fake, if any?"[36]

Williamson agonized over nature faking. Meanwhile Frank Hurley had other worries related to the legitimacy of calling *Pearls and Savages* an "ethnographic" film. Yet the fact that Frank Hurley photographed "underwater" scenes of tropical fish and corals in an aquarium also put him squarely in the category of "nature faker." Perhaps the fact that he had also gone on expedition to the home of tropical fish and corals, carrying an aquarium, lessened the issue. Hurley, though, seemed to sense the importance of remaining vague on the topic. While it was a matter of practicality that he went to the tropics to photograph fish and corals with an aquarium, he never spoke directly about the intervention of the aquarium or made mention of it in the book *Pearls and Savages*. Instead, the public, including the London audience in 1924, assumed he had "broken entirely new ground."[37] Captions encouraged readers to imagine that Hurley was underwater with the fish. The label to one photograph in the *Illustrated London News* read, "Recalling the 'zoo' aquarium," thereby simultaneously acknowledging and denying the fish were photographed in a tank.[38] Slippery language was important because, as Jason E. Hill and Vanessa R. Schwartz point out about news pictures of the day, the picture's promise was "to render a transparent account of the reported event."[39]

If Frank Hurley had called himself an "artist" rather than a documentarist, the problem of realism, truth, and authenticity might have been superfluous, since artists have license to consciously and unconsciously explore the threshold between reality and fiction. But Hurley chose to work in reportage as a documenter, which is why, in the First World War, when he was deployed by the Australian Imperial Forces as an official war photographer to record the event, an incident involving combination printing to create more dramatic images brought him into disrepute. He was accused of producing a war record that was "nothing short of fake."[40] In 1924, in New York City, the question of authenticity was raised again.

When Hurley traveled to New York in 1924 to promote *Pearls and Savages* (retitled for the American market as *The Lost Tribe*), he presented a synchronized lecture and film screening at Carnegie Hall. The *New York Times* reported that the illustrated lecture was well attended, and one point of interest was the way Hurley and his colleagues were worshipped in New Guinea by "savage natives who believed them to be gods."[41] There was considerable buildup of Hurley's event. The event, however, was introduced by Carl Akeley.

Akeley reacted to the claim that Hurley's film *The Lost Tribe* was an "ethnographic" study of Torres Strait Island and Papuan peoples in their island and sea environments. Scientists had started to question the appropriateness of filmmaking and visual imagery for anthropological purposes. Was cinema a legitimate ethnographic tool? Alison Griffiths has observed how a paradigmatic shift took place in anthropology in the first two decades of the twentieth century, away from "visual forms of data collection in favor of the production of written texts."[42] In science, the written text came to surpass the visual text for truth and accuracy. The image was increasingly thought unreliable and untrustworthy and less responsive than the notebook to the everyday lives of others. One outcome was a concern among museum scientists to sharply demarcate films for science created by scientists, and films for science created by nonscientists. The former were pure and true and uncontaminated by art and entertainment. However, Gregg Mitman makes an important point when he argues that the history of film developed outside the hallowed space of the museum, and for this reason film "could never escape its entertainment role," even in the hands of scientists.[43]

A review of Hurley's synchronized lecture performance in New York carried the headline "Hurley's Pictures at Carnegie Hall: Akeley Slams Producers." Even before the lecture performance for *The Lost Tribe*, Akeley criticized promoters and film directors for succumbing to sensationalism over scientific exploration, discovery, and education: "Before enjoying the picture the audience had to listen to an attack on film distributors by Carl E. Akeley, curator of the Museum of Natural History. Mr Akeley dwelt for some time on a charge that distributors impair the scientific merits of travel pictures by editing them solely with an eye to thrills. Then he enlarged his attack to include directors."[44]

While it was important to Akeley that popular amusement did not interfere with science, what audiences wanted from wildlife films and travelogues was drama and theatrics. There was the case of a film by Cherry Kearton titled *Roosevelt in Africa* that was released in 1910. Roosevelt and Akeley had embarked on a hunting and conservation expedition to Africa sponsored by the Smithsonian Institute. They hunted many thousands of animal specimens for taxidermy, and while camera hunting took copious photographs. But Kearton's film failed to excite audiences because it lacked action, which proved that "audiences craved drama over authenticity."[45] A little while later, though, William Selig produced *Hunting Big Game in Africa*, where he re-created, in a Chicago studio, a scene of Roosevelt hunting and shooting a lion. The film was fake, but it was also a great commercial success, and no doubt gave Roosevelt pause to reflect on the moral compromise it presented to the higher values he held in relation to authenticity and truth. Mitman expands on the issues and debates by discussing a series of wildlife adventures in Africa that were filmed in Los Angeles by William N. Selig in the 1910s:

> The difficulty in distinguishing between wholesome education and bawdy entertainment was a problem faced by educators, scientists, and philanthropists who wished to cultivate natural history film for a more serious-minded audience. One could not just dispense with the dramatic, since this would largely eliminate the attendance of the popular theater-going public. Emotional drama was necessary, but the question of whether such drama had been authentically captured in the wild or had been created through artifice in order to elicit thrills and generate mass appeal increasingly became a subject of inquiry and concern.[46]

Contradictions, conflicts, inconsistencies, and double-standards were embedded in explorer culture, which was emphatically prone to exaggeration and hyperbole while seeking serious-minded inquiry. At gatherings at the Explorers Club, members told each other stories of "true tales of modern exploration."[47] Effective storytelling, using rhetorical techniques for building tension and anticipation, was an exciting part of the social exchanges between people in the club. It was a requisite, for example, that every story involve hidden danger in an alien world.[48] The Explorers Club, then, was also a place of amusement and theatre. Members at black-tie dinners were entertained by and educated with "live animal displays and exotic menus of scorpion, sea snakes, and other rarities."[49] Rare and exotic animals, dead or alive, added spice and drama to true stories of explorations in remote countries, and symbolized the triumph of man over nature. But members were emphatic about drawing a line between the seriousness of high cultural pursuits and popular, everyday activities.

Robert Peck explains how belonging to that prestigious club meant proving that work undertaken was "serious exploration as opposed to international travel."[50] The separation of work and play, and of productive scientific enterprise from flippant touristic pleasure and popular appeal was paramount. Travel for recreational purposes was called "tourism," and while many tourists traveled to engage with nature and to practice natural history and popular science, the association with leisure, known also as time in excess of work, drove a conceptual wedge between exploration and travel. The boundaries, though, started to crumble with the intensification of tourism after 1950.

The sign for Hurley and Williamson that the period 1920 to 1950 was an era in transition came when the natural world that they had explored in the early twentieth century became the tourist world of the 1950s. It was a change that had an impact on their image, but more so on the work they produced and their future engagements with coral reefs. By 1950, Hurley and Williamson, once celebrated for attempting to make ground-breaking films and photographs in remote regions of the planet, were displaced by younger men with newer machines and greater exploits to boast about. A different kind of underwater explorer emerged whose every move was captured on color film and by the relatively new medium of television. Jacques-Yves Cousteau (1910–1997) changed the situation for underwater photography and filmmaking forever by using small water-proof cameras and portable aqualung technology that were developments of inventions created in the 1920s by Yves Le Prieur and trialed by the surrealist-biologist Jean Painlevé.[51]

Cousteau was also a member of the Explorers Club. His underwater films were in the tradition of the documentarist-as-explorer. He combined heroics, realism, sentimentality, suspense, drama, and teaching. In 1962, four years before J. E. Williamson died, Cousteau, who was then director of the Oceanographic Museum of marine science in Monaco-Ville, Monaco, issued a stamp for the state of Monaco honoring Williamson, a fellow member of the Explorers Club and a pioneer of underwater exploration. The stamp was also in honor of the global significance of an ocean-oriented future.[52]

CHAPTER 12

Color and Tourism

Across the tropics of both hemispheres, coral structures
stud the blue girdle of encircling oceans

Life, 1954

Two events changed Western perspectives on the
underwater and altered the history of subaquatic
photography: self-contained underwater breathing apparatus (scuba), and the love affair the world developed with
the French naval officer Jacques-Yves Cousteau. Cousteau's
escape underwater from the dull routines of everyday life,
his command of submarine space, and the way his body
expressed a new freedom to move around and below and
above the animals and plants of the sea did more to open
peoples' eyes to the wonders of underwater environments
than all the labors of men and women, scientists and explorers who went before, including Cousteau's associate at
the Explorers Club, J. E. Williamson.

A different attitude to oceans began to emerge in the
popular press of the 1950s. A headline to an article in
the *Daily Mercury*, at Mackay, Queensland, on the eastern coast of Australia, inland from the Great Barrier Reef,
announced in 1953, "Aqualung leads us to rapture of the
sea."[1] The tone was dramatic, emotional, and romantic. It
exhibited a positive disposition toward the depths and the
ocean that signaled a cultural shift in Australia. "Rapture
of the sea" was an evocative phrase borrowed from Cous-

teau, who had, in fact, introduced the world in the 1950s to "the rapture of the deep" to explain the dangerous phenomenon experienced by divers suffering nitrogen narcosis. Cousteau explained it as an altered consciousness in which, in a state of wild elation, divers "simply kept on going down. Others simply remained where they were until they drowned. Anything can happen. Your mind is almost obliterated."[2] But "the rapture of the deep" was a phrase that struck a cord, and it was adopted and adapted in the 1950s in the most positive of ways, endearing people toward the underwater.

When Egon Larsen published *Men under the Sea* in 1955, a pioneering study of the ways and means that human beings, over the centuries, had tried to conquer the underwater, he noted, "We are only now beginning to appreciate the beauty and drama that are waiting for us beneath the waves."[3] But commercial forces were also at work in the 1950s to endear the underwater to the public. And they intensified in the 1960s, when, as Mark Orams observes, Cousteau's use of mass media, especially television, and the television series *The Undersea World of Jacques Cousteau*, coupled with new cameras, raised to unprecedented heights an awareness of the underwater as an environment to enjoy and explore.[4] The shift toward the underwater as a new place to seek leisure is captured by a story in *Time* magazine in 1960: "Around the world and across the nation, swimmers are sinking beneath the surface to fly like angels through an alien realm. This fascinating new playground, alive with beauty and tinged with danger, belongs to the skindiver."[5]

Jacques Cousteau's inventions allowed underwater perception to extend beyond vision, to knowing the sea's depths through the body. At the same point in history as Cousteau's early influence, the philosopher Maurice Merleau-Ponty poetically expressed a desire to know the world through the body and "embrace" it from within as going into "the flesh of the world," an idea expressed in a book titled *The Visible and the Invisible* (1964).[6] Merleau-Ponty developed an understanding of the phenomenology of perception while looking at the bottom of a swimming pool "through the water's thickness," seeing the tiles on the floor of the pool through the "flesh" of reflections and "ripples of sunlight" in the underwater— this according to Alphonso Lingis, who in 1968 translated Merleau-Ponty's book into English.[7] Later, Lingis would take up skin diving at coral reefs and write about the experience and sensations of the body drifting in the underwater. What takes hold in the lived experience, he said, is abandonment of the control of the ego as the body drifts as one object among many. He named the feeling "the rapture of the deep," having appropriated the phrase from Cousteau.[8]

Cousteau led the way in imagining and embodying the undersea. With mobile underwater cameras, he revealed a 360-degree view of underwater space as a living, moving ecological entity. By 1956, he had filmed with Louis Malle the film *The Silent World*, the most important underwater documentary to date for showing the depths in color.[9] These events and developments meant that the work that Frank Hurley and J. E. Williamson had conducted decades earlier was dated and no longer representative of the current state of the art of underwater photography.

But Hurley and Williamson had been game changers in their own way. Their still and moving images had shaped the tropical imaginary. Through magazine illustrations, newspaper articles, travelogues, documentaries, and narrative cinema they helped bring about a transition in popular conception of the seas, peoples, and animals of the tropics. Their fictional films, including Williamson's part in *20,000 Leagues under the Sea* (1916) and *The Mysterious Island* (1929), both filmed in the Bahamas, and Hurley's *Hound of the Deep* (1926) filmed in the Torres Strait and *Jungle Woman* (1926), filmed in Dutch New Guinea, had modernized perceptions of oceans and the underwater by representing it through mass media entertainment.

In the excitement generated by Cousteau, and due to new technologies that greatly expanded the scope of underwater photography, especially photography in color, previous efforts to represent the underwater through early still and moving pictures, and precinematic technologies such as dioramas, were increasingly forgotten. In 1954, Fritz Goro (1901–1986), a science photographer for *Life* magazine, teamed up with Axel Poignant (1906–1986), an Australian natural history photographer and cameraman, and published in *Life* a photo essay of the natural history of the Great Barrier Reef in which it was claimed, "Relatively few people, even Australians, have ever laid eyes on this fabulous coral world. The pictures on these pages are therefore windows opening on one of the most extraordinary and little-known regions of the planet."[10]

Frank Hurley's efforts thirty years earlier to open the same window onto the little-known region of the Great Barrier Reef were seemingly forgotten. By the 1950s, J. E. Williamson's work was also at risk of becoming forgotten, until 1954, when Walt Disney produced a second underwater adaptation of *Twenty Thousand Leagues under the Sea*, shot partly in the Bahamas. When reviewing the Disney film in the *Los Angeles Times*, Louis Berg reminded readers that in 1915, also in the Bahamas, Williamson had pioneered underwater movie making with an adaptation of the same story. Berg proposed that "if Hollywood had any respect for its past or hope for its future, it would name one of its boulevards after John Ernest Williamson."[11] To certain historians of film and popular culture, Williamson stood out as a visionary who had molded the underwater imaginary.

With each new technological development for investigating, photographing, and communicating about the underwater came an increasing development in the industry of marine tourism, especially tropical coral reef tourism. Yet coral reef tourism did not start from ground zero in the 1950s; it had slowly gained in popularity over the course of the twentieth century as audiences were exposed to increasing numbers of photographs, films, dioramas, and illustrations, including imagery produced by Frank Hurley and J. E. Williamson. It can be argued that both men were early shapers of what John Urry calls "the growth of the tourist gaze," which occurs when tourists see and experience place through expectations they have developed from exposure to images, especially photographs and films.[12] In this scenario, tourists come to a place with fantasies and expectations that have been predetermined through photographic images.

By integrating the spectacle of the colonial tropics into popular culture and mass media, Frank Hurley and J. E. Williamson produced readymade images of the Bahamas and the Great Barrier Reef. At first, their influence on tourism was incidental to the more pressing concern with the rarity and newness of underwater photographs and films. But throughout the 1930s, and certainly by the 1950s, Hurley and Williamson both worked directly for the tourism industry, an industry that was also instrumental in the business of nation building through the commercialization of national wilderness.

It was the American tourist market that Bahamian and Australian governments tried to attract. At the Bahamas, an American tourist trade was assured because the islands are so close to the coast of North America. The sunny Bahamas had long been a fashionable winter resort. But the growth of American tourism in Australia was impeded by geographical distance. Yet there was great hope throughout the 1930s that with the right advertising, and with modern infrastructure, especially new resorts, the Great Barrier Reef would also attract Americans. At the Golden Gate International Exposition in San Francisco in 1939, for example, the Australian national pavilion displayed a set of "illuvisions," or modern dioramas, showcasing the coral reefs and tropical fish of the Great Barrier in what was seen as "a vigorous and progressive policy of national advertising."[13] The campaign had limited success, and not until 1954, when Fritz Goro, representing *Life*, arrived in the country to produce a photo essay about the Great Barrier, did the dream of the reef as a popular tourist destination seem a distinct possibility: "[The Queensland Minister in charge of the Tourist Bureau] Mr Riordan said that the interest of the United States in the reef was appreciated as it would inevitably lead to tourist traffic from the States to Australia. The reef would be the magnet which would attract American tourists."[14]

What transpired was increasing national competition between agencies variously claiming it was either the Great Barrier Reef of Australia or the coral reefs of the Bahamas that had the clearest water, the most precious shells, the greatest variety of color and form, and the whitest sand. But the Great Barrier Reef had one quality that Australians knew distinguished it from the Bahamas and the Caribbean, and that was size. There was a sense of national pride in Australia that Fritz Goro's *Life* article had emphasized the immensity of the Great Barrier Reef. It was seen as confirmation of Australia's value to the world, and the local Queensland press quoted the article directly: "The mightiest of these and one of the supreme wonders of the natural world is the Great Barrier Reef, a stupendous rampart of submarine buttress, 1260 miles long and 500 feet high enclosing a water domain of approximately 80,000 square miles off the north-eastern coast of the Australian continent. Beside it the works of man are dwarfed. It is the greatest single edifice ever reared by living creatures on the face of the earth."[15]

The impact of Goro's *Life* article on Australia in 1954 illustrates how the growth of tourism in that decade turned a spotlight on local differences in coral reefs rather than connections. It makes quite a contrast to 1928, when the zoologist William Beebe, working among coral reefs in Haiti, wrote about the planetary connection between both regions. He wrote about "the direct H_2O communication with the Barrier Reef of Australia," and how both tropical regions were part of an interconnected ocean space and a single ecological continuum.[16] But thirty years later, increased commercialization of coral reef environments and increased information flow through mass media relating to the visual spectacle of tropical environments produced national competition. When news reached Australia in 1938 that the Bahamas had placed a picture of its coral reefs on a postage stamp, celebrating the reefs as a natural icon and major asset, it led to comments at home about the need for similar initiatives and regrets that while the most beautiful reefs in the world were at the Great Barrier Reef, the asset was yet to be realized.[17]

Then, in 1939, Williamson turned the photosphere itself into a tourist attraction when the government of the Bahamas authorized it as the world's first undersea post office. Tourists liked the novelty of sending postcards marked with the stamp "Sea Floor, Bahamas." It was pointed out by the *New York Times* that the Bahamas was the only country in the world that produced stamps expressing its national identity through the underwater.[18] Celebrating the underwater on postage stamps became a hallmark of Bahamian identity, and it strengthened connections between nationalism and J. E. Williamson. In 2014, when a stamp was released in commemoration of the "75th Anniversary of The World's First Undersea Post Office," the cover image was a picture of Williamson silhouetted against an underwater scene of corals and sea grasses on the Bahamian seafloor.

To help lure tourists to the Great Barrier Reef, scientists and science books began to emphasize nonscientific subjects, such as travel, fishing, and beauty. Two examples are *Great Barrier Reef* (1950) by marine zoologist William Dakin (1883–1950) and *Wonders of the Great Barrier Reef* (1943) by Theodore C. Roughley (1889–1961), zoologist, president of the Royal Zoological Society of New South Wales (1934–36), superintendent of fisheries of New South Wales, and president of the Great Barrier Reef Game Fish Angling Club (1937). When Roughley traveled to the United States in 1945 to lecture about the Great Barrier Reef, he spoke about a "fisherman's paradise," where the climate is "never too warm," and where "an average catch in two hours consists of about fifty fish."[19] As an angler, he was also a shark hunter, and one of his trophies appears in figure 6.1. Roughley was a scientist, but a scientist in charge of the commercialization of the Great Barrier Reef.

In the 1950s, coral reef tourism and natural history had a symbiotic relationship, and it was usual practice for government tourist organizations to publish books of science, and for scientists to write promotional literature about the localized differences of their regions. Natural history books were therefore hybrids serving knowledge, tourism, leisure, and sport:

science was mixed with fiction, data and facts were combined with the marvelous, and explication blended with romance. By mixing up genres and fields of study, what also came to feature in popular science books, including Dakin's *Great Barrier Reef*, was aesthetics.

The publisher of Dakin's *Great Barrier Reef* was the Australian National Publicity Association, an organization administered by the Australian government's Department of Commerce and Agriculture, which also handled tourist publicity abroad. Dakin's text meanders around the science of coral reef formation, Charles Darwin's theory of subsidence, and questions of beauty.[20] He pauses to remember how he had once rowed in a dinghy on a coral lagoon, and, "looking down at the bottom which was sixty feet at least below me, felt quite frightened because there seemed to be no water at all supporting the dinghy. We seemed in fact to be in mid-air and everything below and about us was just perfect in colour."[21]

The clarity of coral reef waters gave Dakin poetic license to fantasize that he was suspended in air and a desire to communicate to readers the cognitive dissonance and disorientations of space and time that they too would experience if they visited the Great Barrier Reef. Imagining his body suspended in a vertical space that was not terrestrial and yet was strangely familiar was a surreal encounter. Not surprisingly, just a few pages later, Dakin remarked how the bizarre and uncanny visual effects of coral reef environments reminded him of scenes from surrealism. Where once it was the avant-garde that verbalized what was surreal about the aesthetics of coral reefs, by 1950 scientists were sufficiently familiar with early twentieth-century avant-garde art, with artists such as André Breton, and with reproductions of surrealist images such as the coral-like patterns in Max Ernst's decalcomania images to recognize the surreal in nature when they saw it. Decalcomania is a biomorphic texture like the effect achieved by blotting color on a surface with a sponge. It appears to some degree in the cover design chosen for Dakin's book, a vision of the underwater by the artist James Northfield (1887–1973), in which dream and reality combine in a magical way (plate 11). The Australian National Travel Association commissioned Northfield in the 1930s to create promotional posters of the Great Barrier Reef, and Dakin chose one of these for his cover. Northfield's image is a medley of natural forms, flowery corals, dramatically patterned fish, and clear water, and it is sympathetic to the idea of the Great Barrier Reef as a fantasy destination where the bizarre and the beautiful comingle.

Crystal clear water is a primary attraction for coral reef tourism and belongs to the lexicon of images that Krista Thompson, in *An Eye for the Tropics*, describes as "instrumental in creating a visual vocabulary through which to represent the sea."[22] By stressing sheer visibility and optical clarity as almost spiritual qualities belonging to tropical water, scientists, artists, and tourist organizations have been able to blend wonder and curiosity with a philosophical reason to travel: the desire for purification. Tropical water promises to nourish the traveler, mentally and physically.

When William Dakin's book was published in 1950, the undersea was conceived in a variety of ways, but two are distinct: through the distancing metaphor of an aquarium, and through the conception of underwater experience as enveloping and immersive. Dakin was an aquarium thinker who looked upon coral pools on the reefs as "superlative aquaria with little fish and quaint creatures."[23] It is wholly fitting, therefore, that the underwater images of corals and reef fish he published in *Great Barrier Reef* were Frank Hurley's aquarium photographs taken in 1922 when Hurley traveled with Allan McCulloch to Papua along the Great Barrier Reef. One of these images appears in figure 9.1.

Frank Hurley's aquarium photographs were apt illustrations for a book by a scientist who looked at coral pools on natural reefs and saw nature as a display in artificial aquariums. However, when in 1955 *Great Barrier Reef* was reprinted posthumously, the editor published an interesting note that alerted readers to a historical transition between the book's first date of publication in 1950 and its reprinting in 1955. Within that time the paradigm had shifted in representations of the underwater. In five years, ways of thinking about the underwater had transitioned from Dakin's inference that the undersea was a display for the eyes much like an aquarium to the undersea as a visceral experience: "Since this book was first written, the aqualung has been invented and, with its aid, skin-divers all over the world are revealing a completely new field in underwater observation, research and photography. The possibilities for Australia's Great Barrier Reef region are limitless." "Editorial Note," Dakin, *Great Barrier Reef* (1955 reprint), 11.

In the intervening years since Dakin's book was first published, the aqualung had enabled perspectives beneath the surface of the sea that would alter the science, tourism, and popular knowledge of the world's coral reefs. Technologies would overcome previous limitations of human vision. However, the aquarium metaphor remained a dominant framework through which tourists at tropical coral reefs in the 1950s conceived of underwater space. Why? Because aquariums promised safe pleasure, they privilege the ocularcentric tradition of acquiring knowledge, and, like miniature objects, they suggest containment and domestication rather than chaos. Characterizing the sea as a giant aquarium brings it symbolically under containment and, as Stefan Helmreich says, it gives rise to "management regimes . . . predicated on the separation of nature and culture."[24] Aquariums also invite the narcissistic suggestion that nature has been designed for human beings, a point that relates to Timothy Morton's argument that the history of "Nature" often entails the reduction of "nonhuman beings to their aesthetic appearance for humans."[25]

Over the centuries, stories that were told about coral reefs by explorers and sailors depicted them as sublime and destructive. Tourism, however, portrayed coral reefs as beautiful apparitions shimmering beneath the sea's surface and "sparkling [with] rubies, sapphires, amethysts and emeralds."[26] The preciousness of color has always been essential for communicating the allure of the tropics; the more exotic and turquoise, the better, since, as David Picard observes, colors such as turquoise "appeal to the mindset of Western middle and upper-class tourists" seeking magic in other worlds.[27] To get people to travel to coral reef environments, show them clear water tinged with the blue hues of the tropics, shades of color like "Bahama blue," a fashionable shade for clothing before 1939.[28] Still and clear, blue and turquoise and pure: this was the visual allure that tropical waters held, and it dominated dreams of coral island destinations. Expelled from those dreams were murkiness and turbulence, along with darkness and the abyss. Tropical island dreams and reveries repressed the darker depths.

Color, observes Jeffrey Cohen, is a "material impress, an agency, and partner, a thing made of other things through which worlds arrive."[29] Through blues tinted with green, and greens tinted with blue, coral reefs materialized as objects of desire. And as color printing, color film, and color reproduction became better at representing the intensities of coral reef environments, so promotion and advertising became more persuasive. As early as 1923, J. E. Williamson recognized that it was detrimental to business to continue photographing and filming the underwater at Bahamian coral reefs in black-and-white. Visualizations had to be modernized, and the underwater represented on screen in the pinks, blues, and reds of an undersea garden and not, as explained in *Popular Science Monthly*, "as gray masses of beautifully shaped plant and tree formations."[30] Color almost superseded form in the effort to depict the tropics as romantic. At the Great Barrier Reef, the material substance of color was also central to the growth of tourism, specifically the "emerald and turquoise and amethyst and pearl-tinted waters of the seas of the Barrier, spread out for the delightful eyes of the visitor, like the riches of Aladdin's Cave."[31]

In 1950, Frank Hurley returned to photograph and film the Great Barrier Reef, commissioned by tourist and travel organizations to photograph holiday destinations such as Heron Island at the reef's southern end. The images he took included a fashion model in a bathing suit at low tide, the same model displaying a tropical fish on the end of a spear, a giant grouper strung up on a wharf at Thursday Island with the fisherman who caught it standing alongside, a Melanesian man unearthing turtle eggs in a sand dune—but nothing, it seems, of the underwater. Instead, Hurley's work in the 1950s focused above sea level and emphasized glamor, sport, and the staging of authenticity in the spectacle of "native" locals presented as a primitive force within the environment.

Hurley then published *Australia: A Camera Study* (1955).[32] For once, the reefs were shown in color—in fact, the most intense colors imaginable (see plate 12). Alongside were descriptions of "tepid waters of crystal sapphire," and "silver sand amongst the coral beds."[33] But they were passages written thirty years earlier, and are identical to descriptions published in 1924 in *Pearls and Savages*, descriptions that were themselves based on earlier diary entries written in 1921 and 1922. By 1955, they had little or no connection to the objects they described. For instance, in 1924, the vision of "tepid waters of crystal sapphire" and "silver sand amongst the coral beds" referred to the quality of water and colors of reefs at Dauko Island, Papua. But in 1955, the same descriptions were applied to Heron Island, a sandy cay off the coast of Queensland, and a very different physical and political location from Papua. Recycling old descriptions and photographs in new domains was a quick and easy working method. It was doubtful whether anyone would detect the repetitions, or the switch of geographies. By applying standardized descriptions to different contexts, Hurley showed himself to be at ease with the tactics and methods of mass culture, because, as Michael Saler says about the modern imagination, it involved being "comfortable with the artifices of mass culture, and the phantasmagoria of symbols and representations that accompany a capitalist economic order."[34]

When Hurley published *Australia: A Camera Study*, the idea was to capture the attention of international readers and potential tourists. In 1921
and 1922, though, when he set forth as a young explorer, the idea was to
discover new experiences, new peoples, new animals and plants, and new
geographies. It's true that his ambitions in 1921 were also commercial, but
the romantic life of the explorer had equal, if not more, weight. By 1955,
however, tourism had tipped the balance. The tourist gaze, writes John
Urry, is something formed by professionals, including photographers,
who construct "visually extraordinary" signs to be gazed upon—visual
signs that are beyond the everyday. These signs act as metaphors, sub-
stituting some feature or detail for the phenomenon itself.[35] In plate 12,
Hurley's psychedelic colors and bewitching forms transform a fragment
of a coral reef into a fantasy, a dream, an otherworldly hallucination of the
Great Barrier Reef itself, and the message it conveys is that the reef will
astound, will transcend the mundane and routine spheres of the everyday
that tourists want to leave behind. From 1921 to 1955, Hurley had produced
coral reefs and the underwater as a modern spectacle, first as an explorer
with an aquarium, and later as a tourist using color photography.

But however eye-catching, Hurley's coral reef photography in the 1950s
looked dated compared with underwater visions that appeared in the same
decade in photographs, magazine illustrations, and films. A new type of
tourist emerged, attracted by skin diving and portable underwater cameras,
seduced by advertising images that visualized the underwater through
"hedonism, exhilaration and exoticism."[36] In 1951, color photographs of
the Great Barrier Reef, taken underwater, started to appear.[37] With underwater filmmaking and photography developing in tandem with coral reef
tourism, it was, as Susan Sontag explained "positively unnatural to travel
for pleasure without taking a camera along."[38]

Western tourists saw it as their right to consume, occupy, and appro-
priate coral reef environments, and Mimi Sheller's observation about the
Caribbean also applies to the Great Barrier Reef: these were places "open
to be invaded."[39] Historically, they were Britain's "colonial possessions,"
and despite differences between colonies, the difference that most Europeans were concerned with was the collective difference of the colonies to
Britain. Specificities of place were subordinated to the view that colonies
were, collectively, sites "of the exotic, of adventure and exploitation."[40] Not
many people gave a thought to the traditional owners of the Great Barrier
Reef or their long history of caretaking of the underwater zone that en-

circled the islands of the reef. Yet the tourism industry depended on the labor of Aboriginal and Torres Strait Islander peoples. Indigenous peoples provided labor for tourism, but as Celmara Pocock explains, Indigenous Australians were "largely invisible" in the promotion of tourism, something she attributes to the state of "colonial race relations."[41] Eurocentricism in tourism meant that the region of the Great Barrier Reef, as well as the islands of the Bahamas, were seen symbolically "not in the west" but "of the west."[42] In the Western worldview, places regarded as colonial possession and their "objects," including native peoples, landscapes, flora and fauna, were directed to the desires, products, and concerns of the West. Promoted in the West as paradise, these places also became "destinations," especially places to play.

By the 1950s, both the Bahamas and the Great Barrier Reef were advertised as fantasy destinations for the Western body to lose itself to the rapture of clear turquoise and emerald water. Two advertisements (plates 13 and 14) for the Bahamas and the Great Barrier illustrate how tropical water became synonymous with relaxation, liberation, and freedom. "Diving to pleasure" is how one newspaper article in Australia in 1953 described the new emphasis on undersea tourism.[43] Plates 13 and 14 also show how the leisured body in the tropics is a horizontal body, not a vertical one, but a body ready to swim and explore the water in three dimensions, and to see and think the underwater from within it rather than from the fringes of reef shores. In response to the fantasy that the tropical underwater is a transformative medium, advertisements made bodies appear elongated and streamlined. Women were transformed into mermaids and sea nymphs who could seemingly live just as easily in air as in water. The possibility of becoming one with nature underpins each vision. Images are by turns glamorous and seductive and promise days of aimless drifting on the surface of a calm, clear blue sea or playing like fish in the underwater.

"Bahamas Tourist Business Booming," read a headline in the *Chicago Daily Tribune* in 1959, in an article written for Americans seeking to experience "a world apart from what we know."[44] The underwater had become such a glamorous commodity in the Bahamas of the 1950s that the Fort Montagu hotel built a swimming pool walled with glass for patrons to observe swimmers from below the water's surface, like looking into a human aquarium.[45] Not only did the glass-walled swimming pool enhance the voyeurism of looking underwater, it made it safe. Always in the back of the tourist mind, whether at the Great Barrier Reef or the Bahamas, was the thought that peacefulness was superficial in light of the savagery and dangers that were said to be lurking beneath the surface. At the Bahamas and the Great Barrier Reef, in the back of the tourist mind was the thought of sharks.

It was commonly known in the Bahamas that J. E. Williamson was a shark fighter, and through his influence the reputation of Bahamian waters as shark-infested became entrenched. The scene was therefore set for the Bahamas to become a preferred location for shark action movies. One relatively recent production was *Jaws IV: The Revenge* (1987), directed by Joseph Sargent. But in 1965, Terrence Yong directed *Thunderball*, a story about the British secret agent James Bond. The *New York Times* noted about *Thunderball* that for a film about a superhero, "the scenery in the Bahamas is an irresistible lure."[46] Tropical, exotic, relaxed, liquid blue, this was the setting in which James Bond (played by Sean Connery) conducted business. Jeremy Black describes the character of James Bond as a colonial adventurer upholding the empire, and in the context of the Bahamas at the time of its independence from Britain, the figure of James Bond struck just such a figure.[47] But *Thunderball* also featured Bond as a futuristic figure in scuba gear submerged and embattled underwater, surrounded by enemies, human and shark. On shore, when a villain throws someone in a swimming pool on New Providence Island, and sharks are released into the pool through a trap door, a fatal attack ensues. With this scene in which the savagery of nature, in the form of the sharks enters the safety of culture, in the form of the pool, the audience experiences a line being crossed. The swimming pool scene in *Thunderball* facilitates the return of the repressed, whereby the fear of the inhospitable sea that modernity claimed to have conquered through technology returns. The audience is reminded that the ocean is, as Steve Mentz observes, "always outside us; it is the place on the earth that remains inimical to human life."[48]

In the 1920s, Frank Hurley and J. E. Williamson had promised to conquer the underwater and domesticate it by capturing it in images. But neither Hurley nor Williamson appears to have taken advantage of the new technology that emerged in the 1950s to photograph and film marine animals in their habitats with underwater cameras. Like explorers in the 1920s, explorers in the 1950s also celebrated the idea of overcoming and dominating the sea. As one newspaper said about Cousteau: with "an aqualung on his back, and flippers on his feet, Cousteau was able to glide and course beneath the water with fish and other piscatorial creatures around and below and above him."[49] People applauded the idea that the aqualung, with the freedom it gave humans to roam and explore, spy, and survey the oceans, had transformed the sublime and intimidating sea into what a press article in 1956 intimated was an aquarium: "If fish hate men, the man they have reason to hate most is a French naval officer named Jacques-Yves Cousteau. He turned the world's seas into a goldfish bowl by co-inventing the aqualung in 1943."[50]

At last, many thought, Man has penetrated the depths of the sea and taken control. With the sea conquered, imaginations ran riot on the potential of oceans as a resource. In the 1950s, the sea was once again a frontier, but less for adventure than for intense commercial exploitation. The tone of writing in the 1950s was futuristic. *Mechanics Today*, for example, fantasized about a modern "Sea City" with underwater buildings designed to breathe air. The dream, claimed the author, was coming closer every day because "modern science and industry have launched a joint endeavor to conquer the sea, as they have the land and air."[51] But developing in parallel with increasing elation at the domestication and exploitation of the sea was an environmental conscience warning about the fragility of oceans. It was expressed most famously in the 1950s by Rachel Carson in *The Sea around Us* (1951), and later echoed by Sylvia A. Earle in *The World Is Blue* (2009). However, no one thought the sea was in real danger from human exploitation: it was too big and too full of life.

So much was promised for the future commercialization of the Great Barrier Reef when *Life* photographer Fritz Goro put a camera underwater in 1954 and photographed a school of blue damselfish swimming over staghorn coral. In 1954, damselfish were the most plentiful species at the Great Barrier Reef. But today, due to coral bleaching and human-induced climate change, their future seems threatened. The whiteness of bleached corals in reefs that are dying makes such a strong contrast to their blue bodies that damselfish are more easily preyed upon than when corals are alive and full of color.[52] What they need for the future is an environment they can hide in, a phenomenon that Frank Hurley remarked upon during his visit to the Great Barrier Reef in 1921: "Neither order nor rule controlled the color scheme of this profound camouflage. It was as if a million seeds of strange plants had been thrown together, and had sprung into plants blossoming in utter confusion. Drifting on a waveless sea, and peering through a pane—three fathoms of crystal water—this sublime spectacle lay glorified."[53] Today damselfish need the "profound camouflage" of a healthy coral reef; a profusion and abundance of forms and colors; a protective, vivid, habitat where animal, vegetable, and mineral are difficult to distinguish and where even the most dramatically colored species can hide in plain sight and not stand out like specimens in an aquarium.

PART V

The Great Acceleration

CHAPTER 13

The Anthropocene

The profound camouflage that impressed Frank Hurley at the Great Barrier Reef when he looked through the surface of the sea to a coral reef below was the same effect that struck J. E. Williamson about coral reefs at the Bahamas:

> The brilliant colours of tropical fish are truly a source of wonder; but now as you visit the floor of the ocean, you can observe these gaudy creatures in their natural haunts and understand why they are so highly coloured. Here the corals, sea fans, sponges and other marine growths fairly glow with colour. Scarlet, crimson, rose-pink, lilac, orange, brilliant yellow, vivid greens, blues of every shade, blend and intermingle. Against this background, swimming between the corals and sponges, the most brilliant of tropical fish blend perfectly with their surroundings and change their colours to suit the environment of the moment.[1]

Descriptions of colors are a useful way of understanding how Hurley and Williamson conceived the nonhuman world. When speaking about the impact on themselves of reef animal colors, they said they felt enchanted by corals, anemones, sponges, and fishes, and were also excited by the affective dimension of nature and how it invigorated the human mind by producing wonder. In this conceptualization, bright colored things were considered inherently beautiful, and their power to evoke disinterested delight similar to a work of art. But when speaking about the relevance of colors to reef animals, and in explaining the value of color for anemones, fishes, sponges, and corals, Hurley and Williamson concluded that colors for animals were functional attributes related to camouflage and survival. The possibility that the colors of reef animals were also connected with aesthetic pleasure and enchantment for the animals themselves did not enter the realm of possibility.

The question of color, therefore, and of how colors of coral reefs were theorized by Hurley and Williamson, is an example of what Jane Bennett calls "the narcissism of humans in charge of the world."[2] What they denied to the animals of the reefs, they granted to the human: the capacity for reason over instinct and aesthetic pleasure over functional mechanisms. In 1921, Frank Hurley was enchanted by the sublime spectacle that lay before him at the Great Barrier Reef. But the state of enchantment, argues Bennett, can point in two different directions: "The first toward the humans who *feel* enchanted and whose agentic capacities may be thereby strengthened, and the second towards the agency of the things that *produce* (helpful, harmful) effects in human and other bodies."[3]

Affect, as Bennett goes on to say, is not specific to human bodies. In the context of a history in which "nature" has been conceived as a passive stage on which humans act out their lives, the affective power of corals has mostly been considered in relation to humankind. Consequently, the way Hurley and Williamson responded to the natural environment was to demarcate human and nonhuman animals on the basis of consciousness. It is precisely the kind of response that Bennett, in *Vibrant Matter: A Political Ecology of Things* (2010), seeks to deconstruct and challenge in order to recalibrate relations between the human and nonhuman. To that end, Bennett invites readers to "picture an ontological field without any unequivocal demarcations between human, animal, vegetable, or mineral."[4]

In modernity, the coral reef was a new stage on which to make history, and when any part of the planet is conceived as a stage for human players to move across, it is also conceived as a background for them, and therefore as an object. But in the twenty-first century, as Clive Hamilton argues, the tables have turned: "The stage on which we make [history] has now entered into the play as a dynamic and capricious force."[5] In the "Anthropocene"— a new geological era in which "humankind has emerged as the most powerful influence on global ecology"—the concept of the coral reef as a stage, an object, and source of resources has inverted and become the coral reef as an active agent in the threat of planetary demise.[6]

Corals are not a force of nature in which nature is conceived as something beyond the human but a powerful agent and participant in the creation of a "political ecology" in which humans and nonhumans are today entangled not separate.[7] But in the histories told in *Coral Empire* about the coral reefs of the Bahamas and the Great Barrier Reef and how media culture and mass reproduction produced the coral reef as a modern spectacle, the relations between humans and corals was always to human advantage. Industry, culture, education, and knowledge all benefited. Over time, though, coral reef ecosystems of the Pacific and Atlantic suffered ongoing damage, and this now affects human worlds too: the loss through global warming of coral reef ecologies and the destruction of vivid, teaming, kaleidoscopic underwater worlds that had so captivated the human imagination in the early twentieth century is a threat to all life, including human life.

There is little or no evidence that Hurley and Williamson ever felt they were part of, and implicated in, the lives of vast populations of animals and plants that colonized the coral reefs they explored. There are few if any instances when they expressed feeling entangled, existentially, with coral reef animals going about their daily lives. On the contrary, the separation of human and nonhuman animal was absolute and based on subject-object distinctions. What often makes Williamson seem cold and cruel arises from traits that prevailed in modern Western history: unswerving belief in human exceptionalism and in animal object-hood. When a stingray lay dying before him on the deck of a boat, Williamson noticed, "There is something human about the face of a ray, and when we got it on deck this one seemed to be trying to say something—trying to utter some articulate words. With frightened eyes, its expression was almost pathetic."[8]

Almost human in expression, but an object of pity, Williamson's account of the dying stingray recalls a recent argument put forward by Timothy Morton about human pity toward neighbors in the nonhuman world, in which he states that "pity for the living world is an aspect of a sadistic relish for devouring it."[9] When nature is externalized as something outside and distant to human life, pity is another way of reinforcing the exclusion and objectification of others. Through Williamson's writing, people learned to fear, pity, condescend to, objectify, and even violate the animals of Bahamian reefs. And visually, through the distancing barrier of the photosphere's glass, the glass walls of still-life dioramas he helped construct, and the spectacle of the cinema screen, Williamson's audiences also came to understand the underwater as an alien and distant entity.

The actions and aspirations of Frank Hurley and J. E. Williamson tell us a great deal about human and animal relationships in the past. We know that Hurley and Williamson endeavored, through photography and filmmaking, to make the unknown world of the underwater known to people who never ventured out of their own immediate surroundings. Through photography and filmmaking, they shaped the idea of wilderness according to their own desires and used media channels to distribute images far and wide that influenced thinking about oceans and the underwater by inducing wonderment and fear. They were future-focused: the underwater was a new frontier of adventure, beauty, knowledge, and resourcefulness. But as Peter Ellyard writes in *Designing 2050* (2008), "The difficulty is that the future that some may so enthusiastically endeavour to 'make' (according to their own image) may well deprive others of the future they desire."[10]

When Frank Hurley put living corals in an aquarium in Papua in 1922 to create an artificial reef to photograph and film so that audiences could enjoy the spectacle, the corals bleached and died and the aquarium repeatedly became a tomb. Over time, so many people took coral stock as if the supply were endless that it was necessary to protect the future of the Great Barrier Reef by making it a marine park. By 1923, coralline lime for fertilizer, cement, and industry was actively sought in the reef.[11] Mining would continue at the Great Barrier Reef for forty years, until in 1981 the poet, Judith Wright, and the conservation movement brought about the reef's World Heritage listing.[12] Similarly, the American Museum Expedition in 1927 and the Williamson–Field Museum expedition in 1929 stripped tons of corals from the underwater—even using dynamite. Only a matter of decades later, the decline in Bahamian coral reefs and Caribbean reefs was so concerning that intervention was needed in the form of conservation programs.[13]

A failing in empathy and foresight concerning oceans and the undersea in the modern period also explains why it was said that the oceans offered limitless stores of wonders and useful things. It was a message and viewpoint reinforced in all corners of life, including literature. In *Twenty Thousand Leagues under the Sea*, for example, the scientist Aronnax found it inconceivable that human beings, destructive though they are, would ever have an impact on the boundlessness of the sea: "Captain Nemo pointed to this prodigious heap of shellfish, and I saw that these mines were genuinely inexhaustible, since nature's creative powers are greater than man's destructive instincts."[14]

Yet the acceleration of the impact of fishing, mining, and tourism meant that oceans and coral reefs were not inexhaustible. In 1955, Egon Larsen called the undersea "a vast, unknown world which we can explore."[15] But within fifty years the ocean went from the mysterious other to an endangered realm, prompting Jean-Jacques Mantello in 2003 to direct a 3-D documentary titled *Ocean Wonderland*, in which audiences were given an immersive experience beneath the surface of the sea at coral reefs in Australia and the Bahamas to raise awareness of anthropogenic damage.

In response to the urgency of planetary, environmental health, and the "Great Acceleration" of anthropogenic impact on Earth in the twenty-first century, art historians Alan C. Braddock and Christoph Irmscher argue that scholars in their field have an ethical responsibility toward the reading of visual images. They call for the discipline of art history to change. They specifically ask art historians to undertake "a more probing and pointedly ethical integration of visual analysis, cultural interpretation, and environmental history than has so far existed in the field."[16] Since Braddock and Irmscher wrote this, other art historians have taken similar positions. Susan Ballard, for example, calls for a revisiting of art history's narratives and argues that "until very recently, art history's engagement with human-animal relationships has been through a romantic and representational lens that is no longer suitable for a time of crisis."[17]

Braddock's, Irmscher's, and Ballard's advice to integrate ethics and visual analysis into environmental history has played a part in shaping the trajectory of chapters in this book. While *Coral Empire* is an investigation of visual culture surrounding the figure of the coral reef in colonial modernity, and focuses on decades that precede the planetary challenges faced today by climate change, it is a history that is very much tied to the present. The present ecological crisis is linked to the time Frank Hurley and Allan McCulloch went on a scientific and picture-making expedition in Papua and scraped away the living polyps from coral stock before bleaching them in the sun so that their films and photographs would not be spoilt by the problem of cloudy water when the corals, under stress, released threads of zooxanthellae. The crisis is also connected to the years when J. E. Williamson and a team of workers at the Bahamas extracted tons of corals and hundreds of fish, and shipped them in the form of bleached skeletons and fish molds to Chicago. The last paragraphs of this last chapter, however, will turn toward examples of contemporary art that lend themselves to the integration of visual analysis and environmental history in relation to the subject of coral reefs in the Anthropocene.

Peter Peryer is an artist who mobilizes our thinking on planetary relationships, not because his work is art about politics but because his work represents a poetics of the world. Peryer's photography serves ecological thinking because it poetically dislocates the viewer from a sense of familiarity with the world. Often his images produce hesitation and uncertainty. Jacques Rancière, in *The Politics of Aesthetics* (2004), argues that art that calls itself political "cannot work," whereas art that operates through hesitation, cognitive dissonance, or the shock of the uncanny will produce in the spectator a sense of defamiliarization that facilitates political engagement.[18] It is in this uncanny space, where humor also plays a part, that Peryer's photographs enable us to negotiate often overwhelming concepts such as the Anthropocene.

A photograph by Peter Peryer dated 2013, seen in plate 15, is simply titled "Coral Reef," the artist avoiding any geographical specificity, and the caption suggesting there is no real reef to which the photograph refers. This is apt, given that the object photographed is an artificial coral reef, although its man-made, engineered nature may not be readily apparent at first glance, since the types of fish and species of coral look authentic. And if there is a delay in recognizing the artificiality of the reef in Peryer's image, it may also be due to the immediate seduction of the apparition before us, which fulfills everything we want a coral reef to be: vivid, teeming, a profound camouflage, kaleidoscopic, healthy, and vibrant. It's the look that tourists go in search of but often never find, except in technologically enhanced postcards. It's the color spectacle described in words by Frank Hurley and J. E. Williamson. It is also the optical effect—the stylized realism—of digital animation screen spectacles of coral reefs underwater of which the eco-oriented *Finding Nemo* (Pixar, 2003) and *Finding Dory* (Pixar, 2016) are two examples. The affective power of those two animated films is present in Peryer's photograph in the hyperreal textures, colors, and pattern effects of coral reefs underwater.[19]

Peryer's photograph persuades us to question our way of thinking about nature in the same way a bouquet of artificial flowers does, or, for that matter, a Frank Hurley photograph of a coral reef underwater and a

Pixar animation. By studying the image *Coral Reef* we come to two realizations linked to the main arguments of this book. The first is the extent to which visual technologies construct the idea of nature, especially nature as pristine wilderness. The second realization is that it is not the likenesses to nature that makes the image so compelling but its likeness to the image we have in our minds of nature, especially wilderness and the environment of an untouched coral reef. The allure lies in the style of the image. Writing about style, Stuart Ewan draws attention to a prediction that Oliver Wendell Holmes made in 1859 about the reproducible photographic image in which he prophesied that photographs would possess an autonomous life and their own reality. Holmes predicted that photographs would offer "a representation of reality more compelling than reality itself, and—perhaps—even [throw] the very definition of *reality* into question."[20] Putting reality under question goes to the heart of Peryer's photography.

The deconstructive turn of contemporary photography to which Peryer's photograph belongs does not put nature forward as a thing out there to go to, as Timothy Morton characterizes the romantic idea of nature, but as a set of concepts and experiences that have been shaped culturally, including by optical technologies such as dioramas and aquariums. Conceivably, the seductiveness of Peryer's photograph will also shape the way those who view the image will imagine the true appearance of a coral reef and the idea of untouched wilderness.[21] The image asks us to question the conventional idea of nature that defines it as a pure and innocent realm outside human culture and as such shares much in common with contemporary rethinking of the concept of "nature." Morton, for example, argues that "putting something called Nature on a pedestal and admiring it from afar does for the environment what patriarchy does for the figure of Woman. It is a paradoxical act of sadistic admiration."[22] It objectifies the loved object and pronounces it owned, and as a possession renders it voiceless and vulnerable. Stacy Alaimo argues a similar position when she writes that what is problematic is the "opposition between 'human' and 'nature,' since that very opposition undergirds many of our philosophical, ethical, and environmental troubles."[23]

Peryer's photograph estranges us from nature as a thing outside culture and presents us instead with a nonromantic, denaturalized nature. But the image helps us negotiate the ecological impact of the Anthropocene. As T. J. Demos explains in relation to a series of deconstructive photographs of wilderness by Darren Almond, "The strange visuality of these photographs, their seeming artificiality and uncanny quality, betrays, I would suggest, a nature no longer available to us conceptually, experientially, and physically, as it once was. Rather, the images unleash the affects of a nature now outmoded, including the longings and nostalgia for an untouched environment, as well as the critical recognition of the distance and foreclosure of that now outdated mirage."[24]

The term "nature" is a "verbal jack-of-all-trades," according to Arthur O. Lovejoy, writing in 1927 in an article explaining the seventeenth- and eighteenth-century philosophical roots of the emergence of "nature" as a concept. For Williamson and Hurley, "nature" could mean a number of things of the variety outlined by Lovejoy: an empirical reality, something antithetical to man, the outdoors, and an array of objects exterior to the human body for imitation through representation.[25] But for Peryer, the nature we perceive with *Coral Reef* is defined by ironic detachment.

Hyperreal colors and textures, along with perfection in style, rupture the moment by reminding us that it is probably form, rather than content, that we are most attracted to. That sense of detachment is compounded by the way Peryer's photograph objectifies the curious and paradoxical reality of photographs themselves, specifically how distance and framing enable photographs to transform the artificial into the natural. Consequently, the photograph doubles the confusions that the surrealists and André Breton recognized in corals themselves: the lack of clear distinction between the animate and inanimate. Ironic detachment accentuates the superficiality of the image.

For while Peryer's reef is vibrant, it is lifeless; it is teeming but empty. As a photograph of an object whose construction was probably based on photographs of coral reefs, the object in Peryer's image likely has no connection at all to a reef in nature. As a simulacrum, what it suggests is a lost real. At the point of this realization—that the vibrant coral reef in Peryer's photograph may not have any relation at all with an original object in nature—ironic detachment gives way to a sense of melancholy at the loss of an ideal. The artwork exposes the fragility of our dreams, but this, argues Timothy Morton, is exactly why art is such an important platform for illuminating what might otherwise remain buried in the public unconscious. Art, Morton argues, plays a vital role in ecological thinking because it "deals with intensity, shame, abjection, and loss."[26]

Staring back at us from Peryer's photograph is the reality that, increasingly in the twenty-first century, humans are building their own reefs with "super corals" designed to withstand climate change and mitigate the losses of natural reefs due to global warming. In the Bahamas, for example, the practice of inducing corals to colonize artificial reefs has been in operation for much of the late twentieth century.[27] Bringing this detail of environmental history to a visual analysis of Peryer's photograph intensifies the potential of the image to communicate on a level described by Emily Apter as "Planetary Dysphoria"—an aesthetic oriented to "modern day ecological disaster" that pervades contemporary literature, contemporary art, and museum curating.[28]

Kathryn Ferguson, for example, asks a depressing question about the Great Barrier Reef that can apply to all the world's reefs as they undergo unprecedented bleaching. It relates specifically to the primary focus of this book on early mass media and the transformation of coral reefs into modern spectacles. She asks whether we still need the Great Barrier Reef. The reef, after all, is known primarily through images that have grown exponentially in number since the nineteenth century. There is an immense visual record of the reef, should it disappear. But Ferguson poses an even more frightening proposition when she asks whether it isn't the modern spectacle of the avalanche of vibrant photographs and films of coral reefs and the underwater that make us today unable to see and accept "the very real and very pressing reality that what constitutes the beauty of the reef is dying."[29]

Bruno Latour said in 1993 that "the ozone layer is too social and too narrated to be truly natural," and the same is true for coral reefs.[30] Ozone layers and reefs have come into being as much through images in visual and written form as through molecules and gases and coral polyps. It's a point that Iain McCalman makes in his extensive cultural history of the Great Barrier Reef.[31] Peter Peryer's photograph *Coral Reef* is part of the same discussion. So, too, is a recent series of photographs of coral reef dioramas by the artist Frederico Câmara, one example of which is seen in plate 16. Câmara's photograph is not political art, but it engages the political and invites us to follow Alan Braddock's advice to integrate visual analysis, cultural interpretation, and environmental history. The image lays out our ideal expectations of a coral reef, including transparent water and vivid colors. But being part of the deconstructive turn in photography, Câmara's image takes the familiarity of an aquarium diorama, a display designed to create the illusion of reality for visitors, and reveals, rather than conceals, its artifice. He shows how a visit to an aquarium, and photographs of aquariums, have the potential, as previous chapters argue, to determine and shape public perceptions and memories of the true nature of coral reefs. Inside the photograph's frame are signs of the display's construction, and through these signs Câmara's image says two things: this is what a pristine reef looks like, and this is how wilderness is constructed.

Susan G. Davis refers to the carefully manicured world of aquarium displays, specifically those found in the marine park Sea World that opened in San Diego in 1964, as "manufactured marine visions" that select the type of image of the marine world the aquarium, marine park, or museum wants the public to see.[32] Writing about the desire to make nature a spectacle, Davis explains how these popular sites are theatres of entertainment where nature is "othered," objectified, and molded into a predetermined image. This approach to the exhibition of marine nature explains why the coral reef in Câmara's photograph is hypervivid, never bleached, the "tide" always low, leaving the coral growths half exposed instead of hidden, and the varieties of corals dispersed over space to make the display an exciting composition. Yet this image, like Peryer's, is also a momento mori reminding us of the fleeting nature of life, and the deaths of coral reefs happening elsewhere.

Can the past elucidate the present ecological crisis? In the 1970s, Lynn White Jr. was an advocate for the study of the past "to clarify our thinking by looking, in some historical depth, at the presuppositions that underlie [the marriage between] modern technology and science" and its impact on approaches to the environment.[33] Consequently, as White put it, "our ecological crisis is the product of an emerging, entirely novel, democratic culture," and that is why, he continues, we need to "rethink our axioms" that underpin the Western marriage of technology and science.[34] The stories of Hurley and Williamson belong to what White called "the historical roots of our ecological crisis," and which he also asserted is "distinctively Occidental."[35] Today these axioms are being rethought through increasing emphasis on the importance of indigenous ecological knowledges following their repression by Western colonialism.

Williamson and Hurley participated in, and saw themselves as furthering, the writing of popular science. With the exception of Hurley on Dauko Island asking about the purpose of the colors of the reefs, and whether the beauty of corals existed independently of him, they were not much engaged, on a philosophical level, with the environments they photographed, filmed, and wrote about. Hurley's film *Pearls and Savages* looked down on coral reefs from above, and while the camera lingered on the animals of the reef seen in shallow rock pools, the viewpoint encourages the viewer to feel curiosity rather than empathy. In Williamson's work, underwater space and ocean animals are frequently shown as something to fear and combat. Contemporary coral reef filmmaking, by contrast, is motivated by environmental activism and empathy. What does it mean to "become animal" and bridge the divide between human and nature, a divide created by the attitude that nature is something to dominate, study for facts, or react hostilely to?[36]

Contemporary fascination with "crittercam," in which cameras are strapped to animals, is an example of trying to bridge the divide between human and animal by giving agency for representation to the nonhuman. For example, in a recent nature video of the Great Barrier Reef we are given insight into a turtle's perspective. The animal is wired with a GoPro camera set up by scientists wanting to protect turtles at the Great Barrier Reef.[37] This will not appeal to anyone who doesn't like to see wild animals tagged or fitted with cameras. But Donna Haraway has explored the issues surrounding the visualizing practice of "crittercam" as seen on TV on the National Geographic Channel, and her inquiry relates to the constructive way in which it helps us encounter interspecies relations. Haraway concedes it is "hard to specify the positive content of the animals' hermeneutic labor" yet finds an answer in the enmeshing of animals and humans. In her view, "Animals make demands on the humans and their technologies to precisely the same degree that the humans make demands on the animals."[38] Those who find it an intrusion on the purity of nature to mix technology and the wild ocean are reminded that our relationship with the ocean has always been through technology. With the camera, or "techno-remora" as Haraway calls it, strapped to the back of the turtle, it does suggest that humans are in control.[39] Yet when watching the turtle from the viewing position of the GoPro, there is a sense of being drawn into a political frame where much more is at stake than the pleasure of viewing. Rather, there is some hesitation and uncertainty about our own place in the world that comes from seeing the subjectivity and sentience of the turtle. It comes from noticing the turtle's decisions, movements, expressions, actions, thinking, personality, appearance, performance, aesthetics, and subjectivity, and from feeling bound to the turtle through the camera. The crittercam is effective in encouraging something that Jonathan Burt appreciates when watching worthwhile films about animals, namely a "sense of what is meant by our co-habitation with other forms of life."[40]

The successful cohabitation of life forms in an age of climate change has never been more important for coral reefs. The consequences of global warming and human impact on the coral reefs of the Bahamas, the Caribbean, and the Great Barrier Reef were first recorded during periods of mass bleaching in the 1980s.[41] The impact of decades of human exploitation were responsible, and in an attempt to face the problem directly, scientists today recommend managing the transition to climate-changed futures rather than attempting to return the coral reefs at the Great Barrier Reef, and the Caribbean, to former states:

> Reefs in the Caribbean will never resemble the faunal composition of past centuries, owing to the ecological extinction of megafauna such as turtles and manatees, the massive decline of the once dominant branching corals of the genus *Acropora*, the irreversible introduction of the predatory lionfish (*Pterois volitans*) and the ongoing impact of coastal development, overfishing and climate change. Similarly, following a mass-bleaching event and unprecedented mortality in 2016, the corals of the remote northern Great Barrier Reef in Australia are unlikely to have sufficient time to fully recover their former species composition before further major bleaching events occur. Instead of attempting to maintain or restore historical baseline assemblages, the governance and management of coral reefs will need to adapt continuously to the new conditions of the coming centuries.[42]

Given the crisis faced by two "wonders of the world" that are also two vital planetary ecosystems, it is difficult to answer the question of whether looking at the distant history of two modern photographers of coral reefs is relevant to the contemporary world, except to suggest that history illuminates the type of planetary relationships that are needed for the future. Nearly a hundred years ago, Frank Hurley and J. E. Williamson embarked on thrilling expeditions to "capture" nature at the colonial tropics. They framed the ocean with spherical and rectangular glass apertures, changing nature as the Claude Lorrain glass had done in the seventeenth century, by imposing on it a preconditioned way of seeing, and the geometry of culture. Their stories now offer insight into the relationship between human beings and nature, and to the blind spots in human conceptualizations of the sustainability of their interactions with the natural environment.

"The sea is a vast lucky dip," wrote Frank Hurley in 1921 about the animals of the underwater that he regarded as objects.[43] The metaphor of a lucky dip relates to the concept of nature as a mystery that is alien and unknowable. When the sea is imagined as a lucky dip, it is also imagined as a container full of invisible objects that when brought to the surface will bring pleasure and surprise. The ocean thus conceived is like an amusement or a game. There were prizes to take from the sea: surprising, random, entertaining, miscellaneous treasures to seize from what Hurley imagined was a vast aquarium container. The value of the prizes that the sea had to offer lay in their becoming photographic and filmic images for public consumption. The future faced was not one of environmental challenges to coral reefs or to the oceans and the underwater but one of the exciting prospect of integrating photography, filmmaking, and exploration to create sensational images of coral reef environments for entertainment and knowledge.

Many photographs in this book offer glimpses of reefs and animals that have been depleted, polluted, degraded, and ransacked. Many photographs were taken in the Bahamas and the Torres Strait, where human communities as well as coral communities, living on low-lying islands, are now in peril as a result of the effects of global warming, such as increasing sea temperatures and rising water levels, and where people are planning for climate-changed futures.[44] In the 1920s the ocean's resources seemed infinite and inexhaustible, yet today we are urged not "to give up hope for the persistence of Earth's coral reefs."[45]

CONCLUSION

An enigmatic black-and-white photograph in a book by the surrealist André Breton set this project in motion and from an inquiry into its background and provenance, a history of underwater photography and filmmaking at tropical coral reefs unfolded. Unmasking the photograph activated history, awakened the past, and invited a process of navigation through surprising visual and textual pathways. The underwater photograph published in 1937 in *Mad Love* demanded attention and engaged questions about its very nature, and in this regard characterizes the kinds of images that W. J. T. Mitchell discusses in *What Do Pictures Want?* (2005). Mitchell describes these objects as powerful entities that "seem to come alive and want things."[1] The photograph that this book is based on is just such a force. It entices the viewer to explore why and how the underwater photography and filmmaking of coral reefs in the early twentieth century helped define the experience of being modern.

In the art, science, and popular culture of the early twentieth century, coral reefs and the underwater were objects of myth, knowledge, and entertainment, and, in the emerging sphere of mass media, a stage for a modern spectacle. Through the case studies of Frank Hurley and John Ernest Williamson we learn something about the depth of desire and the intensity of excitement that once accompanied the idea of being first to photograph and film marine life underwater at tropical coral reefs, and, more than that, to be the first to release those profitable images to the public through expanding mass communications networks. Through their stories we also come to understand the bravado, anthropocentricism, Eurocentricism, and self-importance that surrounded the life of the explorer—ego-based traits that many now argue led humankind to the environmental crisis and social conflicts faced today. Hurley's expeditions to the Torres Strait and Papua New Guinea served the symbolic expansion of the British colonial enterprise, while Williamson's adoption of the Bahamas served the expansion of the United States to islands that were also a British colony. The kind of nature they went in search of was entangled with the nature of masculinity, a nature that Timothy Morton characterizes as a "rugged, bleak, masculine Nature [that] defines itself through extreme contrasts. It's outdoorsy, not 'shut in.' It's extraverted, not introverted. It's heterosexual, not homosexual. It's able-bodied."[2]

Illuminated by the circumstances of the desire to explore and conquer the underwater at the Great Barrier Reef and the Bahamas, and to capture the undersea and its animals on film, and by the colonial lifestyles enjoyed by Hurley and Williamson, is the way colonialism implicated not only the domination and objectification of the racial Other but also of the nonhuman environment. The energy that emanates from the underwater photograph in *Mad Love* is connected with the energetic attempts, in the early twentieth century, to claim every space on the planet, and to conquer a realm that European explorers, including Frank Hurley and J. E. Williamson, said was virgin territory, without culture or history. Yet the underwater of tropical islands and coral reefs that Williamson and Hurley went in search of was not undiscovered, as they liked to think, but was, instead, the traditional home for generations of peoples, marine animals, and plants.

In the early twentieth century, Hurley and Williamson set out on sea adventures, enchanted by the sea's beauty and strangeness, intent on revealing the hidden world beneath the surface, and equally interested in the "new knowledge" of science and the "old mysteries" of art.[3] They determined to bring images of beauty and danger to excite the public imagination, and other images of encounters with undersea life forms to serve as instruments for science. As explorers, they found it inconceivable that the human race could live on the planet without mastering the undersea. Pulling back the curtains on the underwater was like working magic; it made the invisible visible. But the bravado about underwater photography and underwater exploration belied the insecurities and shortcomings of sometimes-frail heroes.

In order to stand out among peers, Hurley and Williamson often concealed their working methods: a mechanical octopus, an aquarium in Papua. They photographed in the logic of air, in radiant light, and in dryness, creating illusions of immersion in the underwater to bring to spectators the impression of eye-to-eye views and close encounters with the submarine animals and seascapes of the tropics. Conceiving of the underwater as one vast picture composed of colors and forms, they also pictured the underwater as an aquarium. And as zoologist E. M. Stephenson said about aquarium thinking in 1946, "A good aquarium can give a very fair imitation of sea animals and plants in their natural condition. But, even so, we are still *outside* the picture, not part of it."[4] Hurley and Williamson manipulated images to invent new realities; they switched geographies, collaged the past with the present, changed contexts and meanings, fabricated exciting stories, disguised processes of simulation, and faked the reality of nature as well as the nature of reality.

Surprising, paradoxical, and sometimes absurd, the stories in this book construct a picture of the early twentieth-century history of underwater photography, filmmaking, and exploration as contradictory. Modernity itself was contradictory. For instance, in the 1920s, the type of world that Hurley and Williamson wanted for future generations included oceans that were conquered and no longer alien to European explorers. Yet at the same time they were aspiring to make oceans thoroughly represented and narrated, they also wanted the underwater to retain its romantic mystique and the danger, beauty, and awe associated with wilderness. However, the wilderness they went in search of was shaped by the optical devices of dioramas and aquariums, technologies that were also metaphors for a relationship with nature that promised intimacy and closeness but ensured distance and alienation. Much like visiting an art gallery or a museum, Hurley and Williamson looked at the underwater through glass windows. Pragmatically speaking, immersion was difficult, if not impossible, especially with cameras. But philosophically, immersion was equally as challenging because it represented a very different paradigm to thinking about the body's relationship with the underwater realm. Immersion leaves the body open to direct encounters with the alien and the stranger, encounters that require an open mind about relinquishing control and the ego.

People with a receptive and open mind toward nature in the underwater do not find life undersea at coral reefs alien at all, according to Osha Gray Davidson, a writer on science and the environment, and author of

The Enchanted Braid (1998). In the late twentieth century, Davidson's mind was awakened by experiencing the underwater of a coral reef near Heron Island at the Great Barrier. He wrote how "rather than feeling alien in this exotic world, I was filled with the opposite sensation: I felt completely, if inexplicably, at home, as if I belonged there as much as the fishes or the sea cucumbers or the corals."[5] Gone from Davidson's vocabulary are terrestrial metaphors of coral reefs as gardens, trees, branches, or rainforests. They are replaced by visualizations of teeming populations of living things of which he was but one. The conception of coral reefs in Davidson's account embodies what Stefan Helmreich sees as a transition in thinking from what is essentially a nineteenth-century idea of reef environments as curiosities to a new, visceral perception in the later twentieth century of reefs "inviting immersive and fleshy encounters."[6] It is something Alphonso Lingis extended further when he proposed that human bodies are inhabited by coral reefs. Inviting us to imagine our bodies "teeming with polyps, sponges, gorgonians, and free-swimming macrophages continually stirred by monsoon climates of moist air, blood and biles," Lingis also imagined our bodies remembering the sea, and ourselves as a transspecies form of life.[7]

There are no closed borders between human and other life in Lingis's conception of the body as a coral reef. Nothing like his viewpoint emerges in the way Hurley and Williamson reflected on nature. They conceptualized the coral reefs they explored as potential images. They were suppliers of what Stuart Ewen once referred to, in relation to early twentieth-century media culture, as the creation of "a vast and mobile market in images, such as the world had never before seen."[8] The underwater signaled the potential for new, modern, heroic visual experiences. Writing about the heroism of vision, Susan Sontag captured perfectly the spirit of the ambitions that drove Hurley and Williamson, and also the frailty of endeavors like theirs, when she wrote, "There is a peculiar heroism abroad in the world since the invention of cameras: the heroism of vision. Photography opened up a new model of freelance activity—allowing each person to display a certain unique, avid sensibility. Photographers departed on their cultural and class and scientific safaris, searching for striking images. They would entrap the world, whatever the cost in patience and discomfort, by this active, acquisitive, evaluating, gratuitous modality of vision."[9]

Frank Hurley said he understood "the sociology of crowds," insinuating he had insight into the social and psychological needs of audiences.[10]

Williamson said his aspiration was to pry from the undersea its long-held secrets, intimating he had special powers for revealing sights that were previously unseen.[11] Attuned to the psychology of audiences, Hurley and Williamson sensed a successful future with the sea. To attract attention, and to entertain, they exaggerated and dramatized the image of the underwater. They both made reference to the ocean's "depths" as a place to work and photograph, implying a sublime space, even the abyss. As Margaret Cohen points out, while the question of depth has always been relative, the idea of depth has also been a social force exerting a powerful hold on the Western imagination.[12] Hurley and Williamson exaggerated the scale of depth, and they did so because they were storytellers. Their audiences were thrilled by the terror of the deep. In 1982, John Seelye described this terror as an "instinctive dread of deep and dark waters," and argued that it can be sparked as much by ocean imagery that domesticates ocean wilderness, as typified by the writer Jules Verne, as by primitivistic ocean images typified by the writer Herman Melville. The "deep" is a place of fears, and Seelye argued that through multiple sources of science and art, the depth of the sea has been "sublimated into a psychic knot that will never be untied."[13]

Coral Empire has been an unfolding story about coral reef geographies, the cultural significance of corals, underwater photography, and filmmaking in the early twentieth century at the Bahamas and Australia, the subordination of island peoples, the objectification of marine animals, the role of museums in expanding knowledge, the developing economics of underwater imagery for mass media markets, and two men who wanted to make the underwater one of the greatest photographic and cinematographic sensations of their day. In the twenty-first century, the coral reef is among the most photographed, most fetishized, and most exploited of all oceanic phenomena. But in the early twentieth century, the coral reef and the underwater offered filmmakers and photographers, as well as museum curators, an imaginative new field for experimental visualization. Many of the still and moving images reproduced in these pages are accessible through the worldwide web, yet to see Hurley's films firsthand, and images on 16-millimeter film flickering across the screen, is eerie beyond belief. To see the long dead and fleeting creatures of Hurley's artificial undersea, the once living corals and plants, fish and marine life, projected into the present only intensifies an understanding of Hurley's and Williamson's attraction to the enigma of the sea floor and their pursuit of the spectacle of a coral empire.

NOTES

INTRODUCTION

1. In 1915, the idea of photographing underwater for science was newsworthy and novel and was reported in the papers in Australia, including the State of Victoria. See "Submarine Photography," *Euroa Gazette*, supplement, September 28, 1915, 2, accessed February 10, 2018, http://nla.gov.au/nla.news-article153575997.

2. "How Undersea Pictures Are Made," *Table Talk*, June 4, 1925, 50. *Table Talk* was a weekly magazine published in Melbourne, Australia, between 1885 and 1939.

3. Where the Great Barrier Reef begins, and ends, varies according to writers and eras. In the era covered by this book, and in the early 1950s, it was known as the reef that "stretches for roughly 1200 miles from Papua to just south of the Tropic of Capricorn." See "Barrier Reef Films," *Morning Bulletin* (Rockhampton, Queensland), February 15, 1951, 6. When in Papua, for example, Frank Hurley referred to the reefs of the region as "The Great Barrier."

4. Frank Hurley, Tuesday, January 12, 1921, 93, in "A Private Diary, Capt. Frank Hurley. c/-Kodak. Australasia," diary, October 2, 1920–February 1, 1921, series 1, item 7, Papers of Frank Hurley, MS883, National Library of Australia, Canberra, Australia.

5. Helmreich, *Sounding the Limits of Life*, 52.

6. For a study of Jules Verne and cinema, see Taves, *Hollywood Presents Jules Verne*.

7. Chapter 2 addresses early photographers of the underwater and includes discussion of a pioneer of underwater photography, Louis Boutan.

8. See Pocock, "Entwined Histories," 191–92.

9. For expansion on these two forms of cinema, see Altman, "Lecturer's," 61–79.

10. Sontag, *On Photography*, 7.

11. Crylen, "Cinematic Aquarium," 11. Aquarium thinking is also discussed in Eigen, "On the Screen," 229–51; Eigen, "Dark Space," 90–111; and Brunner, *Ocean at Home*.

12. Frank Hurley, "Beneath the Waves Strange Grotesque Life; Forest of Amazing Foliage," *Sun* (Sydney, New South Wales), November 11, 1921, 9.

13. Mentz and Rojas, "Introduction," 1.

14. M. Cohen, "Underwater Optics," 1.

15. "Imperial phantasmagoria" is from Gilroy, *Against Race*, 139–40.

16. Hobsbawm, *Age of Empire*, 8.

17. Arnold, *Problem of Nature*, 142.

18. See "Pacific Islands Added to the Empire: More 'Lumps of Coral,'" *Advertiser* (Adelaide, South Australia), November 16, 1915, 7.

19. Gilroy, *Against Race*, 55.

20. In a chapter titled "Configurations of Blackness" Toni Morrison refers to the reductive way people are characterized by color. The color distinctions "white" and "black" occur frequently in the historical and contemporary literature that informs this book, along with "Indigenous" and "Islander." More specific terms are "African-Bahamian," "Torres Strait Islander," and "Papuan." See Morrison, *Origin of Others*, 55–74.

21. Pike, "Hurley, James Francis."

22. Crotty, *Making the Australian Male*, 229.

23. Driver, *Geography Militant*, 3.

24. McCalman, *Reef*.

25. Adamowsky, *Mysterious Science of the Sea*; M. Cohen, *Novel and Sea*; Rozwadowski, *Fathoming the Ocean*.

26. Burt, *Animals in Film*.

27. Miller, *Empire and the Animal Body*.

28. Rohman, *Stalking the Subject*.

29. Starosielski, "Beyond Fluidity," 152. The second and third phases embrace the 1950s, when underwater mobility expanded dramatically with undersea technologies designed by Émile Gagnan (1900–1979) and the underwater became better known to general audiences through the adventures and filmmaking of Jacques-Yves Cousteau (1910–1997), and then from the 1960s, when underwater filmmaking is distinguished by environmental issues.

30. This is explained, for example, by Pocock, "Entwined Histories."

31. Hughes et al., "Coral Reefs in the Anthropocene," 82.

32. Hage, *Is Racism an Environmental Threat?*

33. The "inhuman" color of gray is discussed in relation to the environment by Jeffrey Cohen in J. Cohen, "Grey," 270.

CHAPTER 1. CORAL EMPIRE

1. E. H. G., "'Shadow-Catching Engines': A Picture Sorcerer in Papua: 'Pearls and Savages' by Captain Frank Hurley," *Illustrated London News*, October 4, 1924, 626.

2. Beebe, *Beneath Tropic Seas*, 6.

3. "The Great Barrier Reef," *Queensland Times* (Brisbane), April 6, 1875, 4.

4. Cook cited in McCalman, "Turtle War," 7.

5. Bennett, *Vibrant Matter*, viii.

6. Helmreich, *Sounding the Limits of Life*, xv.

7. Darwin, *Structure and Distribution*, 1.

8. The question of empathizing with organisms based on bodily difference, such as size and animation, is discussed in Hayward, "Sensational Jellyfish," 177.

9. Scientists today continue to describe corals this way. See Withers, "Empire Building Colonials."

10. Elleray, "Little Builders," 224.

11. "Coral Islands," *Kalgoorrlie Miner* (Western Australia), December 25, 1923, 2.

12. "A Study of Coral Reefs," *Mortlake Dispatch* (Victoria), July 11, 1917, 4.

13. Helmreich, *Sounding the Limits of Life*, 49.

14. Weber, "Madrepore," 916.

15. Marion Endt-Jones explains the history of coral representations in Endt-Jones, "Introductory Essay," 9.

16. Olcott, *Our Wonderful World*, 3.

17. "Assisted Migration," *Courier* (Brisbane, Queensland), June 25, 1861, 4.

18. "Bits of Wisdom," *Albury Banner and Wodonga Express* (New South Wales), March 13, 1908, 12.

19. "Two Germanies," *Central Queensland Herald*, October 13, 1938, 27.

20. M. W. P, "Brotherhoods and their Origin: Organised Fraternities," *Age* (Melbourne, Victoria), September 5, 1942, 7.

21. "*Empire* Trade for 50 Years," *Newcastle Sun* (New South Wales), August 31, 1929, 4.

22. C. H. N., "The Empire Builders," *Sydney Morning Herald*, May 20, 1905, 8. Another press example in which empire builders and coral builders are analogized is W. M. Hughes, "Labor and the Empire," *Barrier Miner* (Broken Hill, New South Wales), June 17, 1908, 2.

23. "Pacific Islands Added to the Empire: More 'Lumps of Coral,'" *Advertiser* (Adelaide, South Australia), November 16, 1915, 7.

24. R. Coupland, "Bird's-Eye View of the British Empire," *Daily Standard* (Brisbane, Queensland), June 20, 1936, 9.

25. Philip J. Bell, "Builders of the Barrier Reef: Beautiful Queensland Corals," *Sydney Mail* (New South Wales), March 9, 1938, 37.

26. Napier, *On the Barrier Reef*, 194.

27. Napier, *On the Barrier Reef*, 196.

28. Krista Thompson describes the relationship of Caribbean light and photography to the touristic gaze. See Thompson, *An Eye for the Tropics*, 6. And J. E. N. Veron describes the significance of light to photosynthesis, reef building corals, and shallow water in Veron, *A Reef in Time*, 30.

29. Driver, *Geography Militant*, 1.

30. Crylen, "Cinematic Aquarium," 11.

31. Marshall and Johnsen, "Camouflage in Marine Fish," 189–91.

32. Stephenson and Stewart, *Animal Camouflage*, 52.

33. Stephenson and Stewart, *Animal Camouflage*, 52.

34. See Darwin on mimicry in Darwin, *On the Origin of Species*, 181.

35. Frank Hurley, "Neptune's Garden: Among the Coral of the Great Barrier," *Sun* (Sydney), September 19, 1926, 23.

36. Gibbings, *Blue Angels and Whales*.

37. Robert Gibbings quoted in Stephenson and Stewart, *Animal Camouflage*, 52.

38. Marshall and Johnsen, "Camouflage in Marine Fish," 193.

39. Lingis, *Excesses*, 8.

40. Frank Hurley, "The Floor of the Coral Sea: Among the Blue-green Opal," *Sun* (Sydney), February 7, 1926, 17.

41. Breton, "Surrealism and Painting," 1.

42. Porter, "Baudrillard," 1.

43. Wulf, *Invention of Nature*, 127.

44. "How Undersea Pictures Are Made," *Table Talk*, June 4, 1925, 50.

45. Siegfried Kracauer in *Theory of Film*, cited in Von Moltke, "Ruin Cinema," 402.

46. Breton, *Mad Love*, 13.

47. For a discussion of modernism and Western culture's "progressive path towards abstraction," see Huijbens, Costa, and Gugger, "Undoing Iceland?," 46.

CHAPTER 2. MAD LOVE

1. Breton, *Manifesto of Surrealism*, 10.

2. Sekula, *Fish Story*, 51.

3. Breton, "Surrealism and Painting," 1.

4. The history of the Paris office of the *New York Times* is discussed in John G. Morris, "Henri Cartier-Bresson: Artist, Photographer and Friend," *News Photographer*, September 2004, 2, Internet Archive, archived October 15, 2012, accessed December 20, 2017, http://archive.li/UJxNo.

5. *National Geographic* claims that the first underwater color photograph published was in 1926 by William Longley. See "Milestones in Underwater Photography," *National Geographic*, accessed October 10, 2017, http://photography.nationalgeographic.com/photography/photos/milestones-underwater-photography/#/hogfish_1428_600×450.jpg.

6. For example, Méret Oppenheim passed on to Breton a photograph of bees swarming on a bicycle seat. Oppenheim's gift to Breton is noted in Barber, "From 'Familiar,'" 157.

7. The photograph is digitized and included in the resources of the website for the Association Atelier André Breton, www.andrebreton.fr.

8. Virginia Pope, "In an Odd World of Coral and Fishes," *New York Times Magazine*, December 1, 1929, 95.

9. The Field Museum–Williamson Undersea Expedition is described by Williamson, *Twenty Years Under the Sea*, 251–55.

10. Pope, "In an Odd World," 95.

11. Drabble, "Submarine Dreams."

12. Kort, "Arnold Böcklin, Max Ernst," 46.

13. (20,000 Leagues Under the Sea. Dir. Stuart Paton. Universal Film Manufacturing Company, 1916.)

14. Verne, Twenty Thousand Leagues, 838.

15. Adamowsky, Mysterious Science of the Sea, 81.

16. See Shick, "Toward an Aesthetic Marine Biology," 71.

17. Burt, "Art and Science of Marine Life."

18. Fretz, "Surréalisme sous-l'eau," 51.

19. Information on rubber fins from Dugan, Man Explores the Sea, 9–11.

20. Cyborgs and border crossing are discussed in Haraway, Simians, Cyborgs, and Women.

21. Crylen, "Cinematic Aquarium," 103.

22. Both are included in Dugan, Man Explores the Sea.

23. Stanley Parker, "Shrimps in the Hair and a Kipper for Picasso," Table Talk, August 20, 1936, 6.

24. Eigen, "On the Screen," 239.

25. Eigen, "Dark Space," 95.

26. Breton, Manifesto of Surrealism, 10.

27. M. Cohen, "Underwater Optics," 2–3.

28. M. Cohen, "Underwater Optics," 17–8.

29. Martínez, "Souvenir of Undersea Landscapes."

30. Norton, Stars beneath the Sea, 159.

31. Ades, "Photography and the Surrealist Text," 187.

32. Sontag, On Photography, 71.

33. Parkinson, "Emotional Fusion," 268.

34. Marine zoologist William Dakin wrote about this popular association between the deep and devolution in William J. Dakin, "To the Bottom of the Abyss," Sydney Morning Herald, August 30, 1947, 9.

35. Polizzotti, Revolution of the Mind, 403.

36. Van Liere, "On the Brink," 269.

37. See comments about intrauterine memories in Dalí, Secret Life, 27. For discussion of the Dream of Venus see Kachur, Displaying the Marvelous, 138.

38. Breton, Mad Love, 11.

39. Breton, Mad Love, 8.

40. Krauss, "Photographic Conditions," 112.

41. Michelet, Sea, 197.

42. Verne, Twenty Thousand Leagues, 527.

43. See "Hospitable Islands: Rediscovering the Bahamas," Register (Adelaide, South Australia), June 6, 1923, 11; "The Barrier Reef," Queensland Times, March 18,

1933, 2; Diana Rice, "An 'Eighth Wonder': Great Barrier Reef, Australia Wins Tourist Popularity," *New York Times*, August 23, 1936, XX10.

44. Brawley and Dixon, *South Seas*, 179.
45. Stephen, McNamara, and Goad, *Modernism and Australia*, 16.
46. The European portrayal of the Pacific as picturesque and other is discussed in Ryan, *Cartographic Eye*, 54–87.
47. Warlick, "Magic, Alchemy and Surrealist Objects," 18.
48. Warlick, "Magic, Alchemy and Surrealist Objects," 18.
49. Toscano and Kinkle, *Cartographies of the Absolute*, 18–19.
50. In December 1945 Breton held a conference in Haiti on the liberation of man. See Depestre, "André Breton," 229–33.
51. Hammerton, *Our Wonderful World*, 1:342.
52. For an account of Breton's encounter with the Sulka mask, see Jeudy-Ballini, "Bringing Aesthetic Emotion," 61.

CHAPTER 3. WILLIAMSON AND THE PHOTOSPHERE

1. S. J. Perelman paraphrased in Taves, *Hollywood Presents Jules Verne*, 23.
2. The "photosphere" is discussed in Todd, *Astronomy*, 127.
3. "Taking Movies at the Bottom of the Sea," *Popular Mechanics* 22, no. 1 (July 1914): 6.
4. Williamson, *Twenty Years under the Sea*, 118–19.
5. Starosielski, "Beyond Fluidity," 153.
6. Helen M. Rozwadowski and David K. van Keuren argue that "what oceanographers have learned about the ocean has been based almost exclusively on what various technologies, or machines, have taught them"; Rozwadowski and van Keuren, xxiii.
7. Doyle Messeley, "A Modern Neptune," *West Australian* (Western Australia), October 19, 1940, 7.
8. Williamson, *Twenty Years under the Sea*, 167.
9. John Ernest Williamson, "Filming the Denizens of the Deep," *Advertiser* (Adelaide, South Australia), September 30, 1933, 8.
10. Crylen, "Cinematic Aquarium," v.
11. Gibson, *Ecological Approach*, 205.
12. Crylen, "Cinematic Aquarium," 25.
13. Armstrong, *Victorian Glassworlds*, 11.
14. Klein, Review of *The Ocean at Home*, 710.
15. Carl Gregory quoted in Burgess, *Take Me under the Sea*, 182.
16. Evans, "Vehicular Utopias," 102.
17. Adamowsky, *Mysterious Science of the Sea*, 162.
18. Williamson, *Twenty Years under the Sea*, 81.
19. Williamson, *Twenty Years under the Sea*, 29.

20. "Submarine Snapshots: Hidden Life Revealed," *Daily Herald* (Adelaide, South Australia), November 17, 1914, 7.

21. The historical information in this paragraph was cited in Burgess, *Take Me under the Sea*, 163–244.

22. Williamson, *Twenty Years under the Sea*, 38.

23. "Taking Movies," 6.

24. Williamson, *Twenty Years under the Sea*, 47–48.

25. Craton and Saunders, *Islanders in the Stream*, 91.

26. Starosielski, "Beyond Fluidity," 153–54.

27. Taves, "Pioneer under the Sea."

28. Williamson, *Twenty Years under the Sea*, 74–78.

29. Advertisement quoted in Taves, "With Williamson," 61.

30. Williamson, *Twenty Years under the Sea*, 77.

31. Williamson, *Twenty Years under the Sea*, 78.

32. William Beebe was also a pioneer of "underwater" photography. The difference between Beebe and Williamson was explained well in the December 1934 issue of *Popular Science Monthly* magazine in which Williamson was acknowledged as the "pioneer photographer of life beneath the waves" while Beebe was described as a "noted explorer and holder of the world's deep-sea diving record," having descended over 2,500 feet below the surface of the water around Nassau. See "Stalk Sea Monsters in Odd Craft: Mysteries of the Deep May Be Solved by New Invasion of Submarine Caverns," *Popular Science Monthly*, December 1934, 45.

33. Dugan, *Man Explores the Sea*, 150.

34. See Martínez, "Souvenir of Undersea Landscapes."

35. See Martínez, "Souvenir of Undersea Landscapes," 2.

36. Eldredge, "Wet Paint," 120.

37. Marx, *Into the Deep*, 166–67.

38. Dugan, *Man Explores the Sea*, 141.

39. Louis Boutan, "Submarine Photography," *Century Illustrated Monthly Magazine*, May 1898, 43.

40. Eigen, "Dark Space," 91.

41. Adamowsky, *Mysterious Science of the Sea*, 153–54.

42. Dugan, *Man Explores the Sea*, 151.

43. Mitman, "Cinematic Nature," 639.

44. "Not a contradiction in terms" is how Trevor Norton describes men of Williamson's era who were both hunter and naturalist, and he cites the example of Theodore Roosevelt. See Norton, *Stars beneath the Sea*, 57.

45. Starosielski, "Beyond Fluidity," 153.

46. Thompson, *Eye for the Tropics*, 185.

47. Williamson, *Twenty Years under the Sea*, 100.

48. William J. Dakin, "To the Bottom of the Abyss," *Sydney Morning Herald*, August 30, 1947, 9.

49. Williamson, *Twenty Years under the Sea*, 165.

50. "Christmas Search for Loch Ness Monster: Holidaymakers Anxious to Secure Photographs," *Scotsman*, December 26, 1933, 10.

51. Marina Warner, "Here Be Monsters," *New York Review of Books*, December 19, 2013–January 8, 2014, 61.

52. "Stalk Sea Monsters," 45.

53. "Deep Sea Monsters Do Exist, Explorer Believes," *Los Angeles Times*, October 12, 1937, A2.

54. See Goodwin, "Remarks on the Polymorphic Image," 8–9.

55. Herrick, "Walks under the Sea," 944.

56. Albury, *Story of the Bahamas*, 10.

57. At Haiti, looking up from underwater, the surface of the sea appeared to William Beebe as a "ceiling." See Beebe, *Beneath Tropic Seas*, 32. When Beebe's bathysphere was brought to the surface, it made him "unconsciously duck" as if it were a hard surface. See Larsen, *Men under the Sea*, 76.

58. Williamson, "Filming the Denizens of the Deep," 8.

59. Williamson, *Twenty Years under the Sea*, 274–75.

60. Williamson, *Twenty Years under the Sea*, 261.

61. Robins, *Into the Image*, 20.

62. "Spatial penetration" is Anthony Vidler's term, and he argues that it was one of the aims of modernity. See Vidler, *Architectural Uncanny*, 217–27.

63. Dugan, *Man Explores the Sea*, 4–5.

CHAPTER 4. THE FIELD MUSEUM–WILLIAMSON UNDERSEA EXPEDITION

1. Williamson, *Twenty Years under the Sea*, 270.

2. Williamson, *Twenty Years under the Sea*, 302.

3. On animals and afterlives, see Alberti, *Afterlives of Animals*.

4. Williamson, *Twenty Years under the Sea*, 255.

5. John Ernest Williamson to Stanley Field, August 9, 1929, Director's Papers: Expedition Correspondence, box 8, item 1 of 5, p. 1, Field Museum Archives, Chicago.

6. John Ernest Williamson to Stephen C. Simms, February 18, 1929, Field Museum–Williamson Undersea Expedition, Director's Papers: Expedition Correspondence, box 8, item 1 of 5, p. 1, Field Museum Archives, Chicago.

7. John Ernest Williamson to Stephen C. Simms, February 18, 1929, p. 1, Field Museum–Williamson Undersea Expedition.

8. John Ernest Williamson to Stephen C. Simms, February 18, 1929, p. 1, Field Museum–Williamson Undersea Expedition.

9. Williamson and Olcott, *Child of the Deep*, vii.

10. Leon Pray to W. Osgood, February 19, 1929, Field Museum–Williamson Undersea Expedition, Director's Papers: Expedition Correspondence, box 8, item 1 of 5, Field Museum Archives.

11. Hannaway, *Winslow Homer in the Tropics*, 167.

12. Hannaway, *Winslow Homer in the Tropics*, 78.

13. Quinn, *Windows on Nature*, 10.

14. John Ernest Williamson to Stephen C. Simms, April 22, 1929, Field Museum–Williamson Undersea Expedition, Director's Papers: Expedition Correspondence, box 8, item 3 of 5, p. 1, Field Museum Archives, Chicago.

15. Parts of the film, including the shark fight, can be seen in Thanhouser, "In the Tropical Seas (1914)."

16. Williamson, *Twenty Years under the Sea*, 67.

17. Williamson, *Twenty Years under the Sea*, 252.

18. Williamson, *Twenty Years under the Sea*, 255.

19. Rohman, *Stalking the Subject*, 1.

20. Haraway, "Teddy Bear Patriarchy," 24.

21. Williamson, *Twenty Years under the Sea*, 281.

22. Williamson, *Twenty Years under the Sea*, 279.

23. Williamson, *Twenty Years under the Sea*, 279.

24. Alum is used in taxidermy to remove moisture and prevent rotting of the skin or hide. Williamson, *Twenty Years under the Sea*, 277.

25. "Carl Akeley," Field Museum, Chicago, accessed December 20, 2017, https://www.fieldmuseum.org/about/history/carl-akeley.

26. Lewis W. Bernard, foreword to Quinn, *Windows on Nature*, 6.

27. Diorama painter James Perry Wilson quoted in Lewis W. Bernard, foreword to Quinn, *Windows on Nature*, 12.

28. In 1920, the Australian Museum in Sydney planned to build a model of a coral pool with specimens collected from the Great Barrier Reef. See "Our Museum," *Sun* (Sydney, New South Wales), May 30, 1920, 5. In 1928, the Queensland Museum in Brisbane built a coral reef diorama complete with blown glass jellyfish and specimens of corals collected from the Great Barrier Reef. See "Storehouse of Treasures: Queensland Museum Attractions," *Brisbane Courier* (Queensland), December 31, 1932, 12.

29. Williamson discusses the Andros Reef expedition in Williamson, *Twenty Years under the Sea*, 251–52.

30. Bullen, *Idylls of the Sea*, 61–62.

31. Reidy, Kroll, and Conway, *Exploration and Science*, 193.

32. Klein, Review of *The Ocean at Home*, 710.

33. "Storehouse of Treasures," 12.

34. Heber A. Longman, director of Queensland Museum, paraphrased in "Storehouse of Treasures," 12.

35. See Miner, "Diving in Coral."

36. Miller, *Empire and the Animal Body*, 58.

37. Richard Conniff refers to the "mad pursuit" of animals in the title of his book. See Conniff, *Species Seekers*.

38. Field Museum, "The 1893 World's Fair," exhibition wall text, *Field Museum of Natural History*, cited February 10, 2014.

39. "Forests of the Deep: Wonderful Coral Growths of Tree-Like Form off the Bahamas," *Illustrated London News*, November 9, 1929, 802–3.

40. Olcott, *Our Wonderful World*, 13.

41. Norton, *Stars beneath the Sea*, 196.

42. Mark Shwartz, "Scientists Look to the Bahamas as a Model for Coral Reef Conservation," *Stanford News Service*, February 20, 2006, accessed December 20, 2017, http://news.stanford.edu/pr/2006/pr-acorals2-022206.html.

43. Glenn Collins, "Rescuing the Diorama from the Fate of the Dodo: In New Appreciation of Old Technique, Museum Remakes the Sea on Dry Land," *New York Times*, February 3, 2003, accessed December 20, 2017, http://www.nytimes.com/2003/02/03/nyregion/rescuing-diorama-fate-dodo-new-appreciation-old-technique-museum-remakes-sea-dry.html.

44. Williamson, *Twenty Years under the Sea*, 280.

45. The documentary, Williamson, "Under the Sea," is available for public viewing at the Field Museum website. https://archive.org/details/UnderTheSeareel1.

46. Rohman, *Stalking the Subject*, 63.

47. Woods, *Weathering the Storm*.

CHAPTER 5. UNDER THE SEA

1. John Ernest Williamson, "Do Sharks Attack Humans Only When Crazed?," *New York Times*, July 30, 1916, SM9.

2. Gilroy, *After Empire*, 12.

3. Craton and Saunders, *Islanders in the Stream*, 91.

4. Horne, *Negro Comrades*, 98–104.

5. Horne, *Negro Comrades*, 100.

6. Williamson, *Twenty Years under the Sea*, 139.

7. "To Film Fight with Shark: J. E. Williamson off to Bahamas for Undersea Photography," *New York Times*, May 10, 1930, 3.

8. Verne, *Twenty Thousand Leagues*, 574.

9. Rediker, "History from below the Water Line," 293.

10. See Rice, "Food for the Sharks."

11. Rozwadowski, "Oceans: Fusing the History," 445.

12. Miller, *Empire and the Animal Body*, 3.

13. Miller, *Empire and the Animal Body*, 3.

14. Williamson, *Twenty Years under the Sea*, 65.

15. Soper, *What Is Nature?*, 10.

16. Thompson, *Eye for the Tropics*, 103.

17. Williamson, *Twenty Years under the Sea*, 64.

18. D'Arcy, *People of the Sea*, 30.

19. Williamson, *Twenty Years under the Sea*, 64.

20. Thompson, *Eye for the Tropics*, 167.

21. Barnhill, "Colonialism."

22. Williamson, *Twenty Years under the Sea*, 285–86.

23. Burt, "Art and Science of Marine Life," 62.

24. Martin Jay quote cited in Robins, *Into the Image*, 20.

25. Eric Barnouw cited in Bousé, "Wildlife Films," 122.

26. Fanon, *Black Skin, White Masks*, 325.

27. Burt, *Animals in Film*, 47–48.

28. "Undersea Film Afire at Explorers' Dinner," *New York Times*, January 19, 1930, 16.

29. Mordaunt Hall, "J. E. Williamson's Film of Undersea Life Climaxed with a Struggle between Divers and an Octopus," *New York Times*, November 24, 1932, 35.

30. Miller, *Empire and the Animal Body*, 3.

31. Loxley, *Problematic Shores*, 56–57.

32. Taves, "With Williamson," 54.

33. "Under-Sea and War Films," *Sydney Morning Herald*, April 3, 1916, 6.

Epigraph: "Fight between Man and Shark," *World's News* (Sydney), September 5, 1914, 1.

CHAPTER 6. WILLIAMSON IN AUSTRALIA

1. Cronon, "Trouble with Wilderness," 7.

2. These are chapter titles in Corbin, *Lure of the Sea*.

3. John Ernest Williamson, "Weird Denizens of the Ocean's Depths: What Wilkins May See," *Chronicle* (Adelaide, South Australia), May 7, 1931, 70.

4. Also known as Captain Frank Hurley, he had acquired this honorary title during his time as official war photographer in the First World War.

5. John Ernest Williamson, "Taking Moving Pictures at the Bottom of the Ocean," *Leader* (Melbourne, Victoria), September 5, 1914, 50.

6. "Submarine 'Movies,'" *Geelong Advertiser* (Victoria), February 24, 1916, 4.

7. Williamson, "Taking Moving Pictures," 50.

8. "Submarine 'Movies,'" 4.

9. "Empire Theatre," *Goulburn Evening Penny Post* (New South Wales), February 9, 1918, 2.

10. Doyle Moseley, "A Modern Neptune," *West Australian* (Western Australia), October 19, 1940, 7.

11. Petterson, *Cameras into the Wild*, 76–78.

12. Bousé, "Wildlife Films," 125.

13. "Sea Photography: Coral Gardens of the Bahamas," *Telegraph* (Brisbane), February 1, 1934, 7.

14. Bousé, "Wildlife Films," 121.

15. "Submarine Photography: Wonders of the Sea," *Journal* (Adelaide, South Australia), September 13, 1913, 7.

16. "How Undersea Pictures Are Made," *Table Talk*, June 4, 1925, 50.

17. "Submarine Snapshots: Hidden Life Revealed," *Daily Herald* (Adelaide, South Australia), November 17, 1914, 7.

18. ["Submarine Snapshots," 7.].

19. "Wonderful Sea Pictures," *Advertiser* (Adelaide, South Australia), October 27, 1924, 13.

20. Ewen K. Patterson, "Creatures of the Coral," *Sunday Mail* (Brisbane, Queensland), March 15, 1931, 20.

21. "Literary Notes," *Australasian* (Melbourne, Victoria), March 28, 1936, 6.

22. Burt, *Animals in Film*, 85.

23. "Adventures on the Ocean Floor," *World's News* (Sydney, New South Wales), Wednesday May 27, 1931, 16–18.

24. "Weird Denizens of the Ocean's Depths," *Chronicle* (Adelaide, South Australia), May 7, 1931, 70.

25. P. Fremont, "Shark Fishing in the Bahamas," *Morning Bulletin* (Queensland), September 5, 1892, 3.

26. "Hospitable Islands: Rediscovering the Bahamas," *Register* (Adelaide, South Australia), June 6, 1923, 11.

27. On this point, see Veron, *Reef in Time*, 49.

28. Robins et al., "Ongoing Collapse."

29. Winton, *Boy behind the Curtain*, 207.

30. The photograph is by Theodore Roughley, an Australian marine zoologist and one of the first scientists to write a popular book on the natural history of the Great Barrier Reef.

31. Mentioned in a previous chapter, the film had three titles: *Terrors of the Deep*, *Thirty Leagues under the Sea*, and *The Williamson Submarine Expedition*.

32. "The Auditorium: The Williamson Submarine Picture," *Table Talk* (Melbourne, Victoria), February 10, 1916, 25.

33. Burgess, *Take Me under the Sea*, 186.

34. Advertisement for the Williamson Submarine Expedition, *Sydney Morning Herald*, March 27, 1916, 2.

35. "Deep-Sea Wonders: Submarine Pictures at Town Hall," *Daily Herald* (Adelaide, South Australia), June 5, 1916, 5.

36. Williamson, *Twenty Years under the Sea*, 69–71.

37. "Adventures on the Ocean Floor," 17.

38. "Adventures on the Ocean Floor," 16.

39. "Shark Hunt: Islanders Armed with Knives," *Western Star and Roma Advertiser* (Toowoomba, Queensland), March 18, 1922, 5.

40. Gilroy, *Against Race*, 139.

41. Patterson, "Creatures of the Coral," 20.

42. "To Be Photographed," *Central Queensland Herald* (Queensland), October 4, 1934, 15.

43. Saville-Kent, *Great Barrier Reef*, 324–35.

44. "Queensland's Monster: A Gigantic Tortoise: Scientist's Views," *North Western Courier* (Narrabri, New South Wales), October 11, 1934, 7.

45. "'£100 for Holiday Makers': Snare a Sea Serpent," *Courier-Mail* (Queensland), December 9, 1933, 17.

46. Neville De Lacey, "Williamson Will Give Us Real Mermaids on Real Sea-Bed," *Sunday Mail* (Brisbane, Queensland), May 5, 1935, 24.

47. Frank Hurley, "The Floor of the Coral Sea: Among the Blue-green Opal," *Sun* (Sydney), February 7, 1926, 17.

48. Evans, "Vehicular Utopias," 99.

49. Andrew Pike mentioned in Petterson, *Cameras into the Wild*, 107.

50. "Secrets of Deep: Papua's Fastnesses; Science and the Film," *Sun* (Sydney, New South Wales), December 1, 1920, 8.

51. "Secrets of Deep."

52. Frank Hurley, "Floor of the Sea: Sublime Spectacle in Tropics: Australia's Marvelous Beauties; Adventures on a Coral Reef," *Sun* (Sydney, New South Wales), January 16, 1921, 3.

53. Dixon, *Photography, Early Cinema*, 175.

54. David Millar interviewed in Rebecca Latham and Alan Nash, "Out of the Blizzard," transcript, *Australian Story: ABC Television*, June 7, 2001, accessed February 25, 2017, http://www.abc.net.au/austory/transcripts/s305854.htm.

Epigraph: Firebrace, "Aquarius in Question," 30.

CHAPTER 7. HURLEY AND THE FLOOR OF THE SEA

1. This is a sample of the kinds of phrases that accompanied the multitude of articles about Frank Hurley's expeditions and exploits.

2. Millar, *From Snowdrift to Shellfire*, 147.

3. Frank Hurley, "A Great Enterprise: Filming the Great Barrier Reef," *Richmond River Herald and Northern Districts Advertiser* (New South Wales), December 7, 1920, 1.

4. Frank Hurley, "A Private Diary, Capt. Frank Hurley. c/-Kodak. Australasia," diary, October 2, 1920-February 1, 1921, series 1, item 7, pp. 1–2, Papers of Frank Hurley, MS883, National Library of Australia (NLA).

5. Burgess, *Take Me under the Sea*, 159.

6. "Painting Pictures under Ocean Depths," *Collie Mail* (Perth, Western Australia), September 14, 1918, 3.

7. Rozwadowski, "Playing," 162.

8. Hurley, "Great Enterprise," 1.

9. Frank Hurley, "Our Unknown Lands: Lucky Dips under a Wonder Sea," *Sun* (Sydney, New South Wales), February 3, 1921, 9.

10. Dixon, *Photography*, 194.

11. Said, *Culture and Imperialism*, 25.

12. Frank Hurley, "Beneath the Waves Strange Grotesque Life; Forest of Amazing Foliage," *Sun* (Sydney, New South Wales), November 11, 1921, 9.

13. Frank Hurley, caption to a submarine photograph in Hurley, *Pearls and Savages* (1925), 83.

14. Lecture films discussed in Petterson, *Cameras into the Wild*, 39.

15. Frank Hurley quoted in Putnam, foreword to *Pearls and Savages* (1925), iv.

16. Frank Hurley, "Frank Hurley's Diary, When on Tour with the Shackleton Picture, 21 November 1919–20 January 1920," diary, November 21, 1919–January 20, 1920, series 1, item 6, p. 63, Papers of Frank Hurley, NLA MS883.

17. Featherstone, "Heroic Life," 165.

18. "Frank Hurley: The Man Who Made Australia's 'Unknown' Known," *Bundaberg Mail and Burnett Advertiser*, August 31, 1915, 3.

19. Claude Lévi-Strauss cited and quoted in Driver, *Geography Militant*, 1.

20. Frank Hurley, "Private Diary," p. 1, Papers of Frank Hurley, NLA MS883/1/7.

21. Reidy, Kroll, and Conway, *Exploration and Science*, 190.

22. Sharp, *No Ordinary Judgment*, 9.

23. Sharp, *No Ordinary Judgment*, 199.

24. Sykes, "Underwater Photography," 1581.

25. Sykes, "Underwater Photography," 1581.

26. Also in 1928, in Haiti, William Beebe's colleague, filmmaker Floyd Crosbie, managed to secure underwater moving pictures of a coral reef using a watertight movie camera and tripod, something Beebe was very excited about. Beebe himself utilized an array of aquariums transported from New York to Haiti to study corals and fish. See Beebe, *Beneath Tropic Seas*, 200.

27. Information about the Crane Pacific Expedition was taken from Webb, "Official/Unofficial Images."

28. Frank Hurley, "Private Diary," p. 55, Papers of Frank Hurley, NLA MS883/1/7.

29. Frank Hurley, "Private Diary," p. 24, Papers of Frank Hurley, NLA MS883/1/7.

30. Frank Hurley, "Private Diary," pp. 24–26, Papers of Frank Hurley, NLA MS883/1/7.

31. Vesely, "Surrealism, Myth and Modernity," 87.

32. Sontag, *On Photography*, 54–55.

33. Frank Hurley, "Private Diary," pp. 93–94, Papers of Frank Hurley, NLA MS883/1/7.

34. Adamowsky, *Mysterious Science of the Sea*, 133.

35. Frank Hurley, "Private Diary," p. 93, Papers of Frank Hurley, NLA MS883/1/7.

36. Clifford, *Traveling Cultures*, 34–35.

37. Chappell, *Double Ghosts*, 38.

38. Frank Hurley, "The Wondrous Sea Floor," *Richmond River Herald and Northern Districts Advertiser* (New South Wales), February 11, 1921, 3.

39. See "Goldfish at Home," *Mail* (Adelaide, South Australia), July 12, 1930, 12.

40. One in the series was titled *Secrets of Nature—Fathoms Deep beneath the Sea* (1922). It was filmed by H. M. Lomas and edited by William Pycraft. See "Fathoms Deep beneath the Sea (Secrets of Nature Series)" in the British Film Institute, *National Film Archive Catalogue*, 62.

41. Frank Hurley, "Mechanics, My Aid in the Wilds," *Popular Science Monthly* 104, no. 3 (March 1924): 40.

42. See Andrews, *Beyond Adventure*; Lord, *Peary to the Pole*.

43. Hurley, "Great Enterprise," 1.

44. Frank Hurley, from *McClure's* magazine (1924), cited in Dixon, "What Was Travel Writing?," 75.

45. Rozwadowski, "Oceans," 443.

46. Brawley and Dixon, *South Seas*, xiii.

47. Eden and Society for Promoting Christian Knowledge (Great Britain), preface to *Guinea Gold*.

48. Frank Hurley, "Floor of the Sea: Sublime Spectacle in Tropics: Australia's Marvelous Beauties; Adventures on a Coral Reef," *Sun* (Sydney, New South Wales), January 16, 1921, 3.

49. Frank Hurley, "Private Diary," p. 91, Papers of Frank Hurley, NLA MS883/1/7.

50. A diagram is included in Frank Hurley, "Private Diary," p. 91, Papers of Frank Hurley, NLA MS883/1/7.

51. Jay, *Downcast Eyes*, 56.

52. Hillel Schwartz, *Culture of the Copy*, 197.

53. Frank Hurley, "Diary No. 1. My Second Expedition to Papua, 1922," diary, August 29, 1922–October 28, 1922, series 1, item 10, p. 72, Papers of Frank Hurley, NLA MS883.

54. "What the Camera Saw on the Ocean Bed," *Sun* (Sydney, New South Wales), November 4, 1922, 1.

CHAPTER 8. HURLEY AND THE AUSTRALIAN MUSEUM EXPEDITION

1. Saler, "Modernity, Disenchantment," 1.

2. Frank Hurley, "Diary No. 1. My Second Expedition to Papua, 1922," diary, August 29, 1922–October 28, 1922, series 1, item 10, p. 1, Papers of Frank Hurley, MS883, National Library of Australia (NLA).

3. Frank Hurley, "Diary No. 1," p. 13, Papers of Frank Hurley, NLA MS883/1/10.

4. Dixon, *Photography, Early Cinema,* 184.

5. White, "Historical Roots," 18–19.

6. Frank Hurley, "Diary No. 1," p. 1, Papers of Frank Hurley, NLA MS883/1/10.

7. Modernity and transparency is discussed in Vidler, *Architectural Uncanny,* 217.

8. Frank Hurley, November 19, 1922, in "Second Expedition to New Guinea. Diary No. 2, 1922," diary, October 31, 1922–December 4, 1922, series 1, item 11, p. 81, Papers of Frank Hurley, NLA MS883.

9. "Governor's View: Capt. Hurley Criticised," *Daily Mail* (Brisbane, Queensland), February 8, 1923, 7.

10. Frank Hurley, "Diary No. 1," p. 43, Papers of Frank Hurley, NLA MS883/1/10.

11. Petterson, *Cameras into the Wild,* 137.

12. Frank Hurley, "Diary No. 1," p. 46, Papers of Frank Hurley, NLA MS883/1/10.

13. Moore, "Empires of the Coral Sea," 153.

14. Frank Hurley, in "Diary No. B, Frank Hurley's diary, 1921," diary, February 1, 1921–April 30, 1921, series 1, item 8, p. 117, Papers of Frank Hurley, NLA MS883.

15. Grimble, *A Pattern of Islands,* 1.

16. Frank Hurley, "Diary No. 1," p. 46, Papers of Frank Hurley, NLA MS883/1/10.

17. Allan R. McCulloch, "21st–24 Sept 1922," "Journal—New Guinea—1922," journal, AA, no. 1/84, P.079, series 33, item 2, p. 9, University of Sydney Archives.

18. Starosielski, "Beyond Fluidity," 150.

19. "With Their Dusky Friends," *Sun* (Sydney), November 4, 1922, 1. The image is also reproduced as "Hurley and McCulloch on Dauko Island," in McGregor, *Frank Hurley,* 247.

20. Rozwadowski, "Playing," 164–65.

21. Frank Hurley, "Diary No. 1," p. 51, Papers of Frank Hurley, NLA MS883/1/10.

22. Frank Hurley, "Diary No. 1," p. 59, Papers of Frank Hurley, NLA MS883/1/10.

23. Frank Hurley, "Diary No. 1," pp. 74–76, Papers of Frank Hurley, NLA MS883/1/10.

24. Frank Hurley, "Diary No. 1," p. 51, Papers of Frank Hurley, NLA MS883/1/10.

25. Frank Hurley, "My Coral Garden," *Daily Mercury,* November 4, 1922, 7. This article was also published under a different title: Frank Hurley, "A New Guinea Idyll: Extraordinary Garden," *Daily Mail,* November 6, 1922, 7.

26. Dixon, "Prosthetic Imagination," 12.

27. Frank Hurley, "Diary No. 1," pp. 55–57, Papers of Frank Hurley, NLA MS883/1/10.

28. Frank Hurley, "Diary No. 1," pp. 55–57, Papers of Frank Hurley, NLA MS883/1/10.

29. Wijgerde, "Inside a Coral Lab."

30. Allan R. McCulloch, "21st–24th Sept 1922," in "Journal—New Guinea—1922," journal, AA, no. 1/84, P.079, series 33, item 2, p. 12, University of Sydney Archives.

31. Bedford, *Great Barrier Reef.*

32. Bedford, *Great Barrier Reef.*

33. Frank Hurley, "Floor of the Sea: Sublime Spectacle in Tropics: Australia's Mar-

velous Beauties; Adventures on a Coral Reef," *Sun* (Sydney, New South Wales), January 16, 1921, 3.

34. E. H. G., "'Shadow-Catching Engines': A Picture Sorcerer in Papua: 'Pearls and Savages' by Captain Frank Hurley," *Illustrated London News*, October 4, 1924, 626.

35. Specht and Fields, *Frank Hurley in Papua*, 8.

36. McCulloch, "Fishes and the Movies," 104.

37. Letter from Allan McCulloch to Frank Hurley, dated May 5, 1924, cited in Saunders, *Discovery of Australia's Fishes*, 373.

38. Frank Hurley, "Diary No. 1," pp. 55–57, Papers of Frank Hurley, NLA MS883/1/10.

39. Corbin, *Lure of the Sea*, 127.

40. Said, *Culture and Imperialism*, 132.

41. Davis, *Spectacular Nature*, 33.

42. Brunner, *Ocean at Home*, 126.

43. Urry, *Tourist Gaze; Leisure*, 138–39.

CHAPTER 9. PEARLS AND SAVAGES

1. See comments on relations between visual culture and empire in Gilroy, *Against Race*, 139.

2. Barnouw, *Documentary*, 22.

3. Ferguson, "Submerged Realities," 119–20.

4. For a discussion of strangers, alterity and otherness in relation to "home" see Chambers, "Stranger in the House."

5. Frank Hurley, "Diary no B, Frank Hurley's diary, 1921," diary, February 1, 1921–April 30, 1921, series 1, item 8, p. 164, Papers of Frank Hurley, MS883, National Library of Australia (NLA).

6. "Papua and 'The Romany,'" *Register* (Adelaide, South Australia), January 8, 1924, 8.

7. See Brawley and Dixon, *South Seas*, xv.

8. Torre, "Tropical Island," 251.

9. Liz McNiven, "Curator's notes." *National Film and Sound Archives*. Accessed 14 June, 2018. https://aso.gov.au/titles/documentaries.pearls–and–savages/notes/.

10. Dixon, *Photography, Early Cinema*, 170.

11. Frank Hurley, "Diary No. 1. My Second Expedition to Papua, 1922," diary, August 29, 1922–October 28, 1922, series 1, item 10, p. 1, Papers of Frank Hurley, NLA MS883.

12. Newspaper notice of *Pearls and Savages*, *Daily Mail* (Brisbane, Queensland), November 1, 1924, 10.

13. Frank Hurley, "Diary No. 1," pp. 47–48, Papers of Frank Hurley, NLA MS883/1/10.

14. Frank Hurley, "Diary No. 1," 47–48, Papers of Frank Hurley, NLA MS883/1/10.

15. Armstrong, *Victorian Glassworlds*, 11.

16. The history of tourism of the Great Barrier Reef is outlined in McCalman, *Reef*, 275.

17. "At the Movies: Strand Theatre," *Newcastle Sun* (New South Wales), November 9, 1923, 6.

18. "Captain Hurley's New Film," *Sydney Morning Herald*, October 6, 1923, 20.

19. Davis, *Spectacular Nature*, 99.

20. Petterson, *Cameras into the Wild*, 27.

21. Frank Hurley, "Diary No. 1," pp. 74–76, Papers of Frank Hurley, NLA MS883/1/10.

22. Hamera, *Parlor Ponds*, 1–2.

23. Hamera, *Parlor Ponds*, 24.

24. Hurley, *Pearls and Savages*, 105.

25. Hurley, *Pearls and Savages* (1925), 114.

26. Carlson, *Aesthetics and the Environment*, 8.

27. See Morton, *Ecology without Nature*.

28. C. B. Christesen, "Roving the Coral Seas," *Walkabout*, June 1, 1936, 28–31. *Walkabout* was an illustrated magazine published in Australia between 1934 and 1974.

29. Torre, "Tropical Island," 246.

30. Benjamin, "Short History of Photography," 25.

31. Palle B. Petterson mistakenly thinks the underwater scenes in Hurley's film *Pearls and Savages* (1921–24) were taken while he was underwater. See Petterson, *Cameras into the Wild*, 105–6.

32. Frank Hurley, "Neptune's Garden: Among the Coral of the Great Barrier," *Sun* (Sydney), September 19, 1926, 23.

33. Hurley, "Neptune's Garden."

34. Frank Hurley, "Diary No. 1," pp. 55–57, Papers of Frank Hurley, NLA MS883/1/10.

35. Frank Hurley, "Diary No. 1," pp. 53–58, Papers of Frank Hurley, NLA MS883/1/10.

36. Gould, *Leonardo's Mountain*, 65.

37. Allan McCulloch, caption to aquarium photograph of big-eyes (*Apogon norfolcensis*), in McCulloch, "Fishes and the Movies," frontispiece.

38. McCulloch, "Fishes and the Movies," 103.

39. Petterson, *Cameras into the Wild*, 28.

40. Matsuda, *Memory of the Modern*, 169.

41. Allan R. McCulloch, "21st–24th Sept 1922," in "Journal—New Guinea—1922," journal, AA, no. 1/84, P.079, series 33, item 2, p. 12, University of Sydney Archives.

42. Eigen, "Dark Space," 96.

43. McCulloch, "21st–24th Sept 1922," 13.

44. Frank Hurley, "Diary No. 1," pp. 75–76, Papers of Frank Hurley, NLA MS883/1/10.

45. Thayer, "Law Which Underlies Protective Coloration," 126.

46. McCulloch, "21st–24th Sept 1922," 9.

47. Tucker, "Historian," 116.

48. E. H. G., "'Shadow-Catching Engines': A Picture Sorcerer in Papua. 'Pearls and Savages' by Captain Frank Hurley," *Illustrated London News*, October 4, 1924, 622.

49. Frank Hurley, December 21, 1922, "Diary No. 3, 4 December 1922–12 January 1923," diary, December 4, 1922–January 12, 1923, series 1, item 12, p. 40, Papers of Frank Hurley, MS883, National Library of Australia.

50. "The Hound of the Deep," *Armidale Chronicle* (New South Wales), November 27, 1926, 12.

51. Ryan, *Picturing Empire*, 186–87.

52. Frank Hurley, "The Floor of the Coral Sea: Among the Blue-green Opal," *Sun* (Sydney), February 7, 1926, 17.

CHAPTER 10. HURLEY AND THE TORRES STRAIT DIVER

1. Frank Hurley, November 2, "My Diary, 5 November 1914 to 31 December 1915," diary, November 5, 1914–December 31, 1915, series 1, item 2, p. 104, Papers of Frank Hurley, MS883, National Library of Australia (NLA).

2. Frank Hurley, "Thrilling Exploits," *New York Times*, March 9, 1924, X5.

3. Frank Hurley, "Beneath the Waves: Sydney Harbor's Greatest Beauty," *Sun* (Sydney, New South Wales), September 25, 1921, 13.

4. Frank Hurley, "Beneath the Waves Strange Grotesque Life; Forest of Amazing Foliage," *Sun* (Sydney, New South Wales), November 11, 1921, 9.

5. Frank Hurley, "Beneath the Waves Strange Grotesque Life."

6. Norman Friend, "Down Under: Marine Curios," *Cairns Post* (Queensland), December 21, 1925, 14.

7. Frank Hurley, "Neptune's Garden: Among the Coral of the Great Barrier," *Sun* (Sydney), September 19, 1926, 23.

8. Traditional time is discussed in relation to primitivism by Dixon, *Photography, Early Cinema*, 181–86.

9. McNab, *Ghost Ships*, 69.

10. Breton, *Manifestoes of Surrealism*, 9–11.

11. Dixon, *Photography, Early Cinema*, 175.

12. Frank Hurley, "Neptune's Garden."

13. Gilroy, *Against Race*, 39.

14. Gregg Mitman, "The Biology of Peace," cited in Elias, *Camouflage Australia*, 61.

15. William J. Dakin, "Concealment, and Camouflage of the Individual in Warfare," cited in Elias, *Camouflage Australia*, 62–63.

16. McMahon, "Pearl Fishers," 182.

17. "Native Divers," *Uralla Times* (New South Wales), April 29, 1937, 4.

18. Charles Hedley, "Australian Pearl Fisheries," *Australian Museum Magazine* 2, no. 1 (1924): 10. This article contains a photograph of a diver wearing full equipment sitting on deck. The same photograph was first published in Frank Hurley, "With the Pearlers: Down to the Sea Floor," *Sun* (Sydney, New South Wales), April 28, 1921, 3.

19. For example, in the late nineteenth century, Christians were likened to divers searching the dark ocean floor for pearls of wisdom and then returning to the

surface, the sunlight, and God. See "The Pearl Diver," *Maitland Weekly* (New South Wales), September 16, 1899, 13.

20. Frank Hurley, "With the Pearlers: 'Shelling' in Torres Straits," *Sun* (Sydney, New South Wales), April 27, 1921, 2.

21. "The Deep Sea Diver: Perils of his Work," *Daily Mercury* (Mackay, Queensland), December 28, 1926, 8.

22. H. F. Wickham, "The Romance of Pearl Fishing," *Sunday Times*, May 9, 1909, 1.

23. Frank Hurley, "Diary no B, Frank Hurley's diary, 1921," diary, February 1, 1921–April 30, 1921, series 1, item 8, p. 51, Papers of Frank Hurley, NLA MS883.

24. Frank Hurley, "Diary no B," p. 61, Papers of Frank Hurley, NLA MS883/1/8.

25. Frank Hurley, "Diary no B," p. 55, Papers of Frank Hurley, NLA MS883/1/8.

26. Frank Hurley, "With the Pearlers: Land of Topsy-Turveydom and Glorious Undersea Garden," *Sun* (Sydney, New South Wales), April 29, 1921, 2.

27. Compare Hurley, "With the Pearlers: Land of Topsy-Turveydom," and Hurley, "Beneath the Waves Strange Grotesque Life."

28. Hurley, "With the Pearlers."

29. See Frank Hurley, "The Pearl Divers of Thursday Island, North Queensland," *Gilgandra Weekly and Castlereagh* (New South Wales), December 21, 1939, 2.

30. Alain Roger from *Court Traité du Paysage*, quoted and cited in Le Dû-Blayo, "Underwater Landscapes," 122.

31. Le Dû-Blayo, "Underwater Landscapes," 128–29.

32. National Library of Australia, "Pearl Diver Collecting Shells from the Beds of Torres Strait, Queensland," catalogue entry, accessed December 18, 2017, https://nla.gov.au/nla.cat-vn4851320.

33. National Library of Australia, "Pearl Diver."

34. Amanda Hopkinson, "Raymond Grosset," *Guardian*, April 20, 2000, accessed December 20, 2017, https://www.theguardian.com/news/2000/apr/20/guardian obituaries.

35. "The Pearl Diver," *Argus* (Melbourne, Victoria), March 18, 1933, 5.

36. See Frank Hurley, "The Pearl Divers of Torres Strait," *Walkabout*, August 1, 1939, 13–15; Hurley, *Queensland*, 210.

37. Hurley, "Pearl Divers of Torres Strait," 15.

38. See Hurley, "Pearl Divers of Thursday Island," 2.

39. Doordan, "Simulated Seas," 10.

40. D. Schwartz, "Visual Ethnography," 122.

41. Frank Hurley, "The Floor of the Coral Sea: Among the Blue-green Opal," *Sun* (Sydney), February 7, 1926, 17.

42. Torma, "Frontiers of Visibility," 26.

43. Ryan, *Photography and Exploration*, 155.

44. See Jolly, "Australian First World War Photography."

45. Hillel Schwartz, *Culture of the Copy*, 286–87.

46. Walter Benjamin discussed in Leslie, *Walter Benjamin*, 57.

47. From an interview with Lowell Thomas by Walter Cronkite, "Remembering Lowell Thomas," in Plimpton, *As Told at the Explorers Club*, 435.

48. Plimpton, *As Told at the Explorers Club*, dustcover.

49. Conniff, *Species Seekers*, 307–8.

50. Peck, "Explorers Club," 98.

51. Williamson's membership in the Explorers Club from 1928 to the time of his death was verified by the club's archivist, Lacey Flint, in correspondence dated August 25, 2016. Flint also commented that while Hurley was not a member, he "would certainly have qualified."

CHAPTER 11. EXPLORERS AND MODERN MEDIA

1. Marshall Berman, *All That Is Solid Melts Into Air*, 15.

2. See Hillel Schwartz, *Culture of the Copy*, 197.

3. Roosevelt, "Nature Fakers," 258.

4. Kevin Robins discusses the frontier aspects of modernity in Robins, *Into the Image*, 15.

5. Hurley, *Pearls and Savages* (1925), 9.

6. M. Cohen, *Novel and the Sea*, 3.

7. Said, *Culture and Imperialism*, xiii.

8. Featherstone, "Heroic Life," 160.

9. See "Souvenir Programme" for J. E. Williamson's *Beauty and Tragedy under the Sea*, screened in Nassau on September 17, 1981, Director's Papers, box 30, folder 1, Field Museum Archives.

10. See Dixon, *Photography, Early Cinema*, 165; "Lee Keedick Presents Mr. J. E. Williamson . . . Beauty and Tragedy under the Sea," 1932, accessed December 21, 2017, http://digital.lib.uiowa.edu/cdm/ref/collection/tc/id/40248.

11. Thompson, *Eye for the Tropics*, 181.

12. Hammerton, foreword, B1.

13. See Boulenger, "Wonderland," 9.

14. "Giant Clams Trap Sea Divers in Grip of Shells," *Popular Mechanics*, May 1924, 685.

15. See Westell, "Monsters of the Mussel Family," 882.

16. Barnouw, *Documentary*, 21.

17. Mitman, *Reel Nature*, 15.

18. Gunning, "Early Cinema," 10.

19. Barnouw, *Documentary*, 24.

20. Barnouw, *Documentary*, 26.

21. The Akeley camera was one of three motion picture cameras taken by the zoologist William Beebe in 1927 to film the life of a coral reef during a five-month expedition to Haiti undertaken by the Department of Tropic Research of the New

York Zoological Society. It was not used underwater. See Beebe, *Beneath Tropic Seas*, 208.

22. Haraway, "Teddy Bear Patriarchy," 37–38.

23. Herbert Schwartz, "In Brightest Africa," 110.

24. Behr, "Akeley Inside," 51–53.

25. During, *Modern Enchantments*.

26. John Ernest Williamson, "Filming the Denizens of the Deep," *Advertiser* (Adelaide, South Australia), September 30, 1933, 8.

27. For a history of camouflage in relation to Australia, see Elias, *Camouflage Australia*.

28. This is discussed by Nemerov, "Vanishing Americans."

29. Williamson, *Twenty Years under the Sea*, 175.

30. Williamson, *Twenty Years under the Sea*, 179–80.

31. Williamson, *Twenty Years under the Sea*, 167.

32. Soister et al., "20,000 Leagues under the Sea," 590.

33. "Undersea Pictures: How They Are Shot," *Western Mail* (Perth, Western Australia), March 16, 1933, 4.

34. Williamson, *Twenty Years under the Sea*, 171.

35. "Romance under the Sea: Ocean Monsters as Screen Stars," *Argus* (Melbourne, Victoria), February 1, 1936, 8.

36. Williamson, *Twenty Years under the Sea*, 177.

37. "Wonderful Film: Striking Coloured Record of Exploration," *Manchester Dispatch*, November 1, 1924, cited in Frank Hurley "Press Cuttings," series 2, items 30, 34, and 35, folio box 1, Papers of Frank Hurley, MS 883, National Library of Australia.

38. E. H. G., "'Shadow-Catching Engines': A Picture Sorcerer in Papua: 'Pearls and Savages' by Captain Frank Hurley," *Illustrated London News*, October 4, 1924, 626.

39. Hill and Schwartz, "Introduction," 5.

40. Bickel, *In Search of Frank Hurley*, 61.

41. "From the Studio's Paint and Powder to the Theatre's Shadows," *New York Times*, February 24, 1924, X5.

42. Griffiths, *Wondrous Difference*, xxv.

43. Mitman, "Cinematic Nature," 640.

44. "Film Shows Headhunters: Hurley's Pictures at Carnegie Hall—Akeley Slams Producers," *Moving Picture World* 67, no. 1 (March 1, 1924): 34, *Internet Archive*, archived November 12, 2014, accessed December 15, 2017, https://archive.org/stream/movpicwor67movi/movpicwor67movi_djvu.txt.

45. Mitman, *Reel Nature*, 9.

46. Mitman, *Reel Nature*, 10.

47. Find the stories told in 1932 at the Explorers Club in Blossom, *Told at the Explorer's Club*.

48. Sherwood, "All in the Day's Work," 289.

49. Peck, "Explorers Club," 98.

50. Peck, "Explorers Club," 100.

51. See Fretz, "Surréalisme sous-l'eau."

52. Taves, "With Williamson."

CHAPTER 12. COLOR AND TOURISM

1. For an impression of Cousteau's impact, see "Aqualung Leads us to Rapture of the Sea," *Daily Mercury* (Queensland), October 5, 1953, 5.

2. Hal Boyle, "Made Seas into Goldfish Bowl," *Canberra Times* (Australian Capital Territory), October 20, 1956, 6.

3. Larsen, *Men under the Sea*, 11.

4. Orams, *Marine Tourism*, 39.

5. *Time* magazine cited in Osgerby, "Rapture of the Deep," 88.

6. Merleau-Ponty, "Working Notes," 270–71.

7. Lingis, "Translator's Preface," xlvi–xlvii.

8. Lingis, "Rapture of the Deep."

9. Delbourgo, "Underwater-Works," 115.

10. Barnett, "World," 76.

11. Louis Berg, "The Man Who Scooped Disney," *Los Angeles Times*, April 4, 1954, 110.

12. Urry, *Tourist Gaze*, 136.

13. Australian National Travel Association, *Australia: General Information*, brochure (Melbourne: Australian National Travel Association, March 1, 1939).

14. "Barrier Reef Big Tourist Asset," *Maryborough Chronicle* (Queensland, Australia), February 13, 1954, 1.

15. *Life* magazine quoted in "Reef Boost in U.S. Magazine," *Courier-Mail* (Brisbane, Queensland), February 12, 1954, 8.

16. Beebe, *Beneath Tropic Seas*, 79.

17. Jay Walker, "Noted in Passing," *Newcastle Sun* (New South Wales), November 28, 1938, 6.

18. "Nassau's 'Sea Floor' Covers," *New York Times*, March 24, 1940, 122.

19. "Australian Talks of Great Coral Reef in South Pacific," *Stanford Daily* 108, no. 31 (November 5, 1945): 4.

20. For a concise summary on Darwin's theory of coral reefs in the context of a contemporary cultural history of coral, see Sponsel, "Darwin's Theory," 70–73.

21. Dakin, *Great Barrier Reef*, 27–28.

22. Thompson, *Eye for the Tropics*, 181–82.

23. Dakin, *Great Barrier Reef*, 48.

24. Helmreich, *Sounding the Limits of Life*, 99.

25. Morton, "X-Ray," 311.

26. Eucalyptus, "Farm & Garden," *Townsville Daily Bulletin* (Queensland), January 26, 1932, 7.

27. Picard, "Island Magic," 26–27.

28. "Wealth of Color and Contrast," *Daily News* (Perth, Western Australia), September 4, 1937, 18.

29. J. J. Cohen, "Introduction," xvi.
30. "Deep Sea Thrillers in Natural Hues for the Movies," *Popular Science Monthly*, February 1923, 32.
31. E. W. G. Bogner, "The Great Barrier Coral Reef," *Horsham Times* (Victoria), July 14, 1933, 10.
32. Hurley, *Australia*.
33. Hurley, *Pearls and Savages* (1925), 105.
34. Saler, "Modernity," 140.
35. Urry, "Tourist Gaze 'Revisited,'" 173.
36. Osgerby, "Rapture of the Deep," 88–89.
37. "Barrier Reef Films," *Morning Bulletin* (Rockhampton, Queensland), February 15, 1951, 6.
38. Sontag, *On Photography*, 9.
39. Sheller, "Demobilizing," 13.
40. Ashcroft, Griffiths, and Tiffin, *Empire Writes*, 18.
41. Pocock, "Aborigines," 170.
42. Sheller and Urry, "Introduction," 6.
43. "Diving to Pleasure: A New World for All under the Sea," *Age* (Melbourne, Victoria), August 15, 1953, 13.
44. Headline and quote from William W. Yates, "Bahamas Tourist Business Booming to a New Record," *Chicago Daily Tribune*, March 15, 1959, K1.
45. Yates, "Bahamas Tourist," 9.
46. Bosley Crowther, "Screen: 007's Underwater Adventures: Connery Plays Bond in 'Thunderball,'" *New York Times*, December 22, 1965, 23.
47. Black, *Politics of James Bond*, 12.
48. Mentz, *At the Bottom*, 5.
49. N. G. M., "A New World for all Under the Sea," *Age* (Melbourne, Victoria), August 15, 1953, 13.
50. Boyle, "Made Seas into Goldfish Bowl."
51. Floyd B. McKnight, "Can You Live under the Sea?," *Mechanics Today*, November 1953, 77.
52. Coker, Pratchett, and Munday, "Coral Bleaching."
53. Frank Hurley, "Floor of the Sea: Sublime Spectacle in Tropics: Australia's Marvelous Beauties; Adventures on a Coral Reef," *Sun* (Sydney, New South Wales), January 16, 1921, 3.

CHAPTER 13. THE ANTHROPOCENE

1. Williamson, *Twenty Years under the Sea*, 91.
2. Bennett, *Vibrant Matter*, xvi.
3. Bennett, *Vibrant Matter*, xii.
4. Bennett, *Vibrant Matter*, 116.
5. Clive Hamilton, "Climate Change Signals the End of the Social Sciences," *Con-*

versation, January 25, 2013, accessed December 20, 2017, http://theconversation.com/climate-change-signals-the-end-of-the-social-sciences-11722.

6. For definitions and discussion of the Anthropocene and the Great Acceleration, see McNeill and Engelke, *Great Acceleration*, 1.

7. Bennett, *Vibrant Matter*, 112.

8. Williamson, *Twenty Years under the Sea*, 282.

9. Morton, *Ecological Thought*, 75.

10. Peter Ellyard cited in Anthony Judge, "Engendering Viable Global Futures through Hemispheric Integration: A Radical Challenge to Individual Imagination," *Laetus in praesens*, published April 30, 2014, accessed December 17, 2017, https://www.laetusinpraesens.org/docs10s/futoz.php.

11. "A Valuable Possession," *Telegraph* (Brisbane, Queensland), September 27, 1923, 8.

12. McCalman, *Reef*, 276–302.

13. Mark Shwartz, "Scientists Look to the Bahamas as a Model for Coral Reef Conservation," *Stanford News Service*, February 20, 2006, accessed December 20, 2017, http://news.stanford.edu/pr/2006/pr-acorals2-022206.html.

14. Verne, *Twenty Thousand Leagues*, 604.

15. Larsen, *Men under the Sea*, 11.

16. Braddock and Irmscher, "Introduction," 3–4.

17. Ballard, "Stretching Out," 4.

18. Rancière, *Politics of Aesthetics*, 63.

19. David Whitley writes about Pixar's affective environments in Whitley, "Animation, Realism," 152.

20. Oliver Wendell Holmes cited in Ewen, *All Consuming Images*, 24–25.

21. Bate, "Memory of Photography," 256.

22. Morton, *Ecology without Nature*, 5.

23. Alaimo, "Feminist Science," 189.

24. Demos, "Photography," 39.

25. Lovejoy, "'Nature' as Aesthetic Norm," 446.

26. Morton, *Ecological Thought*, 10.

27. For literature on artificial reefs, see Carr and Hixon, "Artificial Reefs," 28; Seaman and Jensen, "Purposes and Practices."

28. Apter, "Planetary Dysphoria," 131.

29. Ferguson, "That's Not a Reef," 232.

30. Latour, *We Have Never Been Modern*, 6.

31. "No mind can take it in . . . so vast is it," writes Iain McCalman in McCalman, *Reef*, 3.

32. Davis, *Spectacular Nature*, 19.

33. White, "Historical Roots," 17.

34. White, "Historical Roots," 19.

35. White, "Historical Roots," 15–16.

36. Becoming animal, an idea proposed in 1980 by Gilles Deleuze and Felix Guat-

tari, challenges humankind to transform attitudes and thinking by imagining new identities that reject distinctions between species but rather impel people toward an indeterminacy of the "human." See Deleuze and Guattari, *Thousand Plateaus*, 232–310.

37. In 2013, scientists Dr. Ian Bell of James Cook University Department of Environment and Heritage Protection, and Christine Hof of the World Wildlife Fund, attached cameras to turtles to track their movements at the Great Barrier Reef. For video evidence and the story, see BBC, "Turtle-Cam Reveals Unique Views of Great Barrier Reef," *BBC News*, uploaded July 5, 2015, accessed December 20, 2017, http://www.bbc.com/news/science-environment-33359951.

38. Haraway, *When Species Meet*, 263.

39. Haraway, *When Species Meet*, 259.

40. Burt, "Art and Science of Marine Life," 49–66.

41. Veron, *A Reef in Time*, 57.

42. Hughes et al., "Coral Reefs in the Anthropocene," 82.

43. Frank Hurley, "Our Unknown Lands: Lucky Dips under a Wonder Sea," *Sun* (Sydney, New South Wales), February 3, 1921, 9.

44. On the Torres Strait and rising tides, see Green et al., "Assessment of Climate Change." On sea levels and the Bahamas, see McGranahan, Balk, and Anderson, "Rising Tide."

45. Hughes et al., "Coral Reefs in the Anthropocene," 88.

CONCLUSION

1. Mitchell, *What Do Pictures Want?*, 9.

2. Morton, *Ecological Thought*, 81.

3. This distinction between the mysteries of art and the knowledge of science in relation to the underwater is found in Eldredge, "Wet Paint," 122.

4. Stephenson and Stewart, *Animal Camouflage*, 43.

5. Davidson, *Enchanted Braid*, 10.

6. Helmreich, *Sounding the Limits of Life*, 49.

7. Lingis, "Animal Body," 167.

8. Ewen, *All Consuming Images*, 25.

9. Sontag, *On Photography*, 89–90.

10. Frank Hurley, "Frank Hurley's Diary, When on Tour with the Shackleton Picture, 21 November 1919–20 January 1920," diary, November 21, 1919–January 20, 1920, series 1, item 6, pp. 34–35, Papers of Frank Hurley, MS883, National Library of Australia.

11. "Prying into Secrets of the Sea," *Observer* (Adelaide, South Australia), April 4, 1925, 62.

12. M. Cohen, "Denotation in Alien Environments," 105.

13. Seelye, "Oceans of Emotion," 207.

BIBLIOGRAPHY

Adamowsky, Natascha. *The Mysterious Science of the Sea, 1775–1943*. London: Routledge, 2015.

Ades, Dawn. "Photography and the Surrealist Text." In *L'amour fou: Photography and Surrealism*, edited by Rosalind Krauss and Jane Livingston, 155–89. Washington, DC: Corcoran Gallery of Art / New York: Abbeville Press, 1985.

Alaimo, Stacy. "Feminist Science Studies and Ecocriticism: Aesthetics and Entanglement in the Deep Sea." In *The Oxford Handbook of Ecocriticism*, edited by Greg Garrard, 188–203. Oxford: Oxford University Press, 2014.

Alberti, Samuel J. M. M., ed. *The Afterlives of Animals: A Museum of Menageries*. Charlottesville: University of Virginia Press, 2011.

Albury, Paul. *The Story of the Bahamas*. London: Macmillan, 1975.

Altman, Rick. "From Lecturer's Prop to Industrial Product." In *Virtual Voyages: Cinema and Travel*, edited by Jeffrey Ruoff, 61–79. Durham, NC: Duke University Press, 2006.

Andrews, Roy Chapman. *Beyond Adventure: The Lives of Three Explorers*. New York: Duell, Sloan and Pearce, 1952.

Apter, Emily. "Planetary Dysphoria." *Third Text*. Vol. 27, Issue 1 (2013): 131–140.

Armstrong, Isobel. *Victorian Glassworlds: Glass Culture and the Imagination, 1830–1880*. Oxford: Oxford University Press, 2008.

Arnold, David. *The Problem of Nature: Environment, Culture, and European Expansion*. Oxford: Blackwell, 1996.

Ashcroft, Bill, Gareth Griffiths, and Helen Tiffin. *The Empire Writes Back: Theory and Practice in Post-colonial Literatures*. 2nd ed. London: Routledge, 2002. Originally published 1989.

Ballard, Susan. "Stretching Out: Species Extinction and Planetary Aesthetics in Contemporary Art." *Australian and New Zealand Journal of Art* 17, no. 1 (2017): 2–16.

Barber, Fionna. "From 'Familiar: Alice Maher' 1995." In *Sources in Irish Art: A Reader*, edited by Fintan Cullen, 150–58. Cork, Ireland: Cork University Press, 2000.

Barnett, Lincoln. "The World We Live In: Part VIII: The Coral Reef." *Life* 36, no. 6 (1954): 74–94.

Barnhill, John. "Colonialism." In *Encyclopedia of World Geography*, edited by R. W. McColl, 191–96. New York: Golson Books, 2005.

Barnouw, Eric. *Documentary: A History of Non-fiction Film*. Oxford: Oxford University Press, 1993.

Bate, David. "The Memory of Photography." *Photographies* 3, no. 2 (2010): 243–57.

Bedford, Randolph. *The Great Barrier Reef.* Sydney, Australia: Art in Australia, 1928.

Beebe, William. *Beneath Tropic Seas: A Record of Diving among the Coral Reefs of Haiti*. New York: G. P. Putnam's, 1928.

Behr, Bernd. "Akeley inside the Elephant: Trajectory of a Taxidermic Image." *Philosophy of Photography* 7, nos. 1 and 2 (2016): 43–61.

Benjamin, Walter. "A Short History of Photography." *Screen* 13, no. 1 (1972): 5–26.

Bennett, Jane. *Vibrant Matter: A Political Ecology of Things*. Durham, NC: Duke University Press, 2010.

Berman, Marshall. *All That Is Solid Melts into Air: The Experience of Modernity*. London: Verso, 1982.

Bickel, Lennard. *In Search of Frank Hurley*. Artarmon, Australia: Macmillan, 1980.

Black, Jeremy. *The Politics of James Bond: From Fleming's Novels to the Big Screen.* Lincoln: University of Nebraska Press, 2005.

Blossom, Frederick, ed. *Told at the Explorer's Club: True Tales of Modern Exploration*. London: George G. Harrap, 1932.

Boulenger, E. G. "The Wonderland beneath the Sea." In Hammerton, *Our Wonderful World*, 1:9.

Bousé, Derek. "Are Wildlife Films Really 'Nature Documentaries'?" *Critical Studies in Mass Communication* 15 (1998): 116–40.

Braddock, Alan C., and Christoph Irmscher. Introduction to *A Keener Perception: Ecocritical Studies in American Art History*, edited by Alan C. Braddock and Christoph Irmscher, 1–23. Tuscaloosa: University of Alabama Press, 2009.

Brawley, Sean, and Chris Dixon. *The South Seas: A Reception History from Daniel Defoe to Dorothy Lamour*. Lanham, MD: Lexington Books, 2015.

Breton, André. *Mad Love*. Translated by Mary Ann Caws. Lincoln: University of Nebraska Press, 1987. Originally published 1937.

Breton, André. *Manifesto of Surrealism*. (1924) In Breton, *Manifestoes of Surrealism*, 1–49. Translated by Richard Seaver and Helen R. Lane. Ann Arbor: University of Michigan Press, 1972.

Breton, André. "Surrealism and Painting." In *Surrealism and Painting*, translated by Simon Watson Taylor, 1–49. London: Macdonald, 1972.

British Film Institute. *National Film Archive Catalogue, Part II: Silent Non-fiction Films, 1895–1934*. Foreword by Sir Arthur Elton. London: British Film Institute, 1960.

Brunner, Bernd. *The Ocean at Home: An Illustrated History of the Aquarium.* New York: Princeton Architectural Press, 2005.

Bullen, Frank T. *Idylls of the Sea and Other Marine Sketches.* London: Thomas Nelson and Sons, 1909.

Burgess, Thomas. *Take Me under the Sea: The Dream Merchants of the Deep.* Salem, OR: Ocean Archives, 1993.

Burt, Jonathan. *Animals in Film.* London: Reaktion, 2002.

Burt, Jonathan. "The Art and Science of Marine Life: Jean Painlevé's *The Seahorse.*" In *Animals and the Cinema: Classifications, Cinephilias, Philosophies,* edited by Sabine Nessel, Winfried Pauleit, and Christine Ruffert, 49–66. Berlin: Bertz + Fischer, 2012.

Carlson, Allen. *Aesthetics and the Environment: The Appreciation of Nature, Art, and Architecture.* London: Routledge, 2000.

Carr, Mark H., and Mark A. Hixon. "Artificial Reefs: The Importance of Comparisons with Natural Reefs." Special Issue on artificial reef management. *Reefs* 22, no. 4 (1997): 28–33.

Chambers, Iain. "A Stranger in the House." In Iain Chambers, *Culture after Humanism: History, Culture, Subjectivity,* 161–83. London: Routledge, 2001.

Chappell, David A. *Double Ghosts: Oceanian Voyagers on Euroamerican Ships.* New York: M. E. Sharpe, 1997.

Clifford, James. *Traveling Cultures: Travel and Translation in the Late Twentieth Century.* Cambridge, MA: Harvard University Press, 1997.

Cohen, Jeffrey Jerome. "Grey." In Cohen, *Prismatic Ecology: Ecotheory beyond Green,* 270–89. Minneapolis: University of Minnesota Press, 2013.

Cohen, Jeffrey Jerome. Introduction to Cohen, *Prismatic Ecology: Ecotheory beyond Green,* xv–xxxvi. Minneapolis: University of Minnesota Press, 2013.

Cohen, Margaret. "Denotation in Alien Environments: The Underwater *Je Ne Sais Quoi.*" *Representations* 125, no. 1 (2014): 103–26.

Cohen, Margaret. *The Novel and the Sea.* Princeton, NJ: Princeton University Press, 2010.

Cohen, Margaret. "Underwater Optics as Symbolic Form." *French Politics, Culture, and Society* 32, no. 3 (winter 2014): 1–23.

Coker, Darren J., Morgan S. Pratchett, and Philip L. Munday. "Coral Bleaching and Habitat Degradation Increase Susceptibility to Predation for Coral-Dwelling Fishes." *Behavioral Ecology* 20, no. 6 (2009): 1204–10.

Conniff, Richard. *The Species Seekers: Heroes, Fools, and the Mad Pursuit of Life on Earth.* New York: W. W. Norton, 2011.

Corbin, Alain. *The Lure of the Sea: The Discovery of the Seaside, 1750–1840.* Translated by Jocelyn Phelps. London: Penguin Books, 1994.

Craton, Michael, and Gail Saunders. *Islanders in the Stream: A History of the Bahamian People.* Vol. 2. Athens: University of Georgia Press, 1998.

Cronon, William. "The Trouble with Wilderness of Getting Back to the Wrong Nature." *Environmental History* 1, no. 1 (1996): 7–28.

Crotty, Martin. *Making the Australian Male: Middle-Class Masculinity, 1870–1920.* Carlton South, Australia: Melbourne University Press, 2001.

Crylen, Jonathan Christopher. "The Cinematic Aquarium: A History of Undersea Film." PhD diss., University of Iowa, 2015. http://ir.uiowa.edu/etd/1839.

Dakin, William. *Great Barrier Reef: And Some Mention of Other Australian Coral Reefs.* Melbourne: Australian National Publicity Association, 1950.

Dakin, William. *Great Barrier Reef: And Some Mention of Other Australian Coral Reefs.* Melbourne: Australian National Publicity Association, 1955.

Dalí, Salvador. *The Secret Life of Salvador Dalí.* New York: Dover Publications, 1993.

D'Arcy, Paul. *The People of the Sea: Environment, Identity, and History in Oceania.* Honolulu: University of Hawai'i Press, 2006.

Darwin, Charles R. *On the Origin of Species.* 6th ed. London: John Murray, 1872.

Darwin, Charles R. *The Structure and Distribution of Coral Reefs.* London: Smith, Elder, 1842.

Davidson, Osha Gray. *The Enchanted Braid: Coming to Terms with Nature on the Coral Reef.* New York: John Wiley, 1998.

Davis, Susan G. *Spectacular Nature: Corporate Culture and the Sea World Experience.* Berkeley: University of California Press, 1997.

Delbourgo, James. "Underwater-Works: Voyages and Visions of the Submarine." *Endeavour* 31, no. 3 (2007): 115–20.

Deleuze, Gilles, and Felix Guattari. *A Thousand Plateaus.* London: Athlone Press, 1980.

Demos, T. J. "Contemporary Art and the Politics of Ecology." In "Contemporary Art and the Politics of Ecology," edited by T. J. Demos. Special issue. *Third Text,* Vol. 27, Issue 1, (2013): 1–9.

Demos, T. J. "Photography at the End of the World: On Darren Almond's *Fullmoon* Series." *Image and Narrative* 16, no. 1 (2015): 32–44.

Depestre, René. "André Breton in Port-au-Prince." In *Refusal of the Shadow: Surrealism and the Caribbean,* edited by Michael Richardson, translated by Krzysztof Fijałkowski and Michael Richardson, 229–33. London: Verso, 1996.

Dixon, Robert. *Photography, Early Cinema, and Colonial Modernity: Frank Hurley's Synchronized Lecture Entertainments.* London: Anthem Press, 2013.

Dixon, Robert. "The Prosthetic Imagination: Frank Hurley and the Ross Smith Flight." *Journal of Australian Studies* 24, no. 66 (2000): 1–22.

Dixon, Robert. "What Was Travel Writing? Frank Hurley and the Media Contexts of Early Twentieth-Century Australian Travel Writing." *Studies in Travel Writing* 11, no. 1 (2007): 59–81.

Doordan, Dennis. "Simulated Seas: Exhibition Design in Contemporary Aquariums." *Design Issues* 11, no. 2 (summer 1995): 3–10.

Drabble, Margaret. "Submarine Dreams: Jules Verne's Twenty Thousand Leagues under the Seas." *New Statesman*, May 8, 2014. Accessed December 14, 2017. http://www.newstatesman.com/2014/04/submarine-dreams.

Driver, Felix. *Geography Militant: Cultures of Exploration and Empire*. Oxford: Blackwell, 2001.

Dugan, James. *Man Explores the Sea: The Story of Undersea Exploration from Earliest Times to Commandant Cousteau*. London: Hamish Hamilton, 1956.

During, Simon. *Modern Enchantments: The Cultural Power of Secular Magic*. Cambridge, MA: Harvard University Press, 2002.

Eden, Charles H., and Society for Promoting Christian Knowledge (Great Britain). Preface to *Guinea Gold, or, The Great Barrier Reef*, 108–9. London: Society for Promoting Christian Knowledge London, 1883.

Eigen, Edward. "Dark Space and the Early Days of Photography as a Medium." *Grey Room* 3 (2001): 90–111.

Eigen, Edward. "On the Screen and in the Water: On Photographically Envisioning the Sea." In *L'architecture, les sciences et la culture de l'histoire au XIXe siècle*, introduction by François Loyer, 229–51. Saint-Étienne: Publications de l'Université Saint-Étienne, 2001.

Eldredge, Charles. "Wet Paint: Herman Melville, Elihu Beder and Artists Undersea." *American Art* 11, no. 2 (summer 1997): 106–35.

Elias, Ann. *Camouflage Australia: Art, Nature, Science, and War*. Sydney, Australia: Sydney University Press, 2011.

Elleray, Michelle. "Little Builders: Coral Insects, Missionary Culture, and the Victorian Child." *Victorian Literature and Culture* 39 (2010): 223–38.

Endt-Jones, Marion. "Introductory Essay." In *Coral: Something Rich and Strange*, edited by Marion Endt-Jones, 9–21. Liverpool, UK: Liverpool University Press, 2013.

Ephemera Collection. National Library of Australia, Canberra, Australia.

Evans, Arthur B. "Vehicular Utopias of Jules Verne." In *Transformations of Utopia: Changing Views of the Perfect Society*, edited by George Slusser, Paul Alkon, Roger Gaillard, and Daniele Chatelain, 99–108. New York: AMS Press, 1999.

Ewen, Stuart. *All Consuming Images: The Politics of Style in Contemporary Culture*. New York: Basic Books, 1984.

Fanon, Frantz. *Black Skin, White Masks*. Translated by Charles L. Markmann. London: MacGibbon and Kee, 1968.

Featherstone, Mike. "The Heroic Life and Everyday Life." *Theory, Culture, and Society* 9 (1992): 158–82.

Ferguson, Kathryn. "Submerged Realities: Shark Documentaries at Depth." *Atenea* 26, no. 1 (2006): 115–29.

Ferguson, Kathryn. "That's Not a Reef. Now *That's* a Reef: A Century of (Re)Placing the Great Barrier Reef." In *Ecosee: Image, Rhetoric, Nature*, edited by Sidney I. Dobrin and Sean Morey, 223–39. Albany: State University of New York Press, 2009.

Field Museum–Williamson Undersea Expedition. Director's Papers: Expedition Correspondence. Field Museum Archives, Chicago.

Firebrace, William. "Aquarius in Question." *Cabinet: A Quarterly of Art and Culture* 48 (winter 2012–13): 30–38.

Fretz, Lauren E. "Surréalisme sous-l'eau: Science and Surrealism in the Early Films and Writings of Jean Painlevé." *Film and History* 40, no. 2 (fall 2010): 45–65.

Gibbings, Robert. *Blue Angels and Whales: A Record of Personal Experiences below and above Water.* London: J. M. Dent and Sons, 1946. Originally published 1938.

Gibson, James J. *The Ecological Approach to Visual Perception.* Boston. Houghton Mifflin, 1979.

Gilroy, Paul. *After Empire: Melancholia or Convivial Culture?* Abingdon, UK: Routledge, 2004.

Gilroy, Paul. *Against Race: Imagining Political Culture beyond the Colour Line.* Cambridge, MA: Belknap Press of Harvard University Press, 2000.

Goodwin, John B. L. "Remarks on the Polymorphic Image." *View* 1 (December–January 1941–2): 8–9.

Gould, Stephen Jay. *Leonardo's Mountain of Clams and the Diet of Worms: Essays on Natural History.* Cambridge, MA: Harvard University Press, 2011.

Green, Donna, Lisa Alexander, Kathy McInnes, John Church, Neville Nicholls, and Neil White. "An Assessment of Climate Change Impacts and Adaptation for the Torres Strait Islands, Australia." *Climatic Change* 102, no. 3–4 (2010): 405–33.

Griffiths, Alison. *Wondrous Difference: Cinema, Anthropology, and Turn-of-the-Century Visual Culture.* New York: Columbia University Press, 2002.

Grimble, Arthur. *A Pattern of Islands.* London: John Murray, 1952.

Gunning, Tom. "Early Cinema and the Variety of Moving Images." *American Art* 22, no. 2 (2008): 9–11.

Hage, Ghassan. *Is Racism an Environmental Threat?* Cambridge, UK: Polity Press, 2017.

Hamera, Judith. *Parlor Ponds: The Cultural Work of the American Home Aquarium, 1850–1970."* Ann Arbor: University of Michigan Press, 2012.

Hammerton, John Alexander, ed. *Our Wonderful World: A Pictorial Account of the Marvels of Nature and Triumphs of Man.* 4 vols. London: Amalgamated Press, 1931.

Hannaway, Patti. *Winslow Homer in the Tropics.* Richmond, VA: Westover, 1973.

Haraway, Donna J. *Simians, Cyborgs, and Women: The Reinvention of Nature.* London: Free Association, 1991.

Haraway, Donna J. "Teddy Bear Patriarchy: Taxidermy in the Garden of Eden, New York City, 1908–1936." *Social Text* 11 (winter 1984–85): 20–64.

Haraway, Donna J. *When Species Meet.* Minneapolis: University of Minnesota Press, 2008.

Hayward, Eva. "Sensational Jellyfish: Aquarium Affects and the Matter of Immersion." *Differences: A Journal of Feminist Cultural Studies* 25, no. 5 (2012): 161–96.

Helmreich, Stefan. *Sounding the Limits of Life: Essays in the Anthropology of Biology and Beyond*. Princeton, NJ: Princeton University Press, 2016.

Herrick, F. H. "Walks under the Sea by a Coral Strand." *American Naturalist* 23, no. 275 (November 1889): 941–56.

Hill, Jason E., and Vanessa R. Schwartz. Introduction to *Getting the Picture: The Visual Culture of the News*, edited by Jason E. Hill and Vanessa R. Schwartz, 1–11. London: Bloomsbury, 2015.

Hobsbawm, Eric John. *The Age of Empire, 1875–1914*. London: Weidenfeld and Nicolson, 1987.

Horne, Gerald. *Negro Comrades of the Crown: African Americans and the British Empire Fight the U.S. before Emancipation*. New York: New York University Press, 2012.

Hughes, Terry P., Michele L. Barnes, David R. Bellwood, Joshua E. Cinner, Graeme S. Cumming, Jeremy B. C. Jackson, Joanie Kleypas, Ingrid A. van de Leemput, Janice M. Lough, Tiffany H. Morrison, Stephen R. Palumbi, Egbert H. van Nes, and Marten Scheffer. "Coral Reefs in the Anthropocene." *Nature* 546 (June 2017): 82–90.

Huijbens, Edward J., Bárbara Maçães Costa, and Harry Gugger. "Undoing Iceland? The Pervasive Nature of the Urban." In *Tourism and the Anthropocene: An Urgent Emerging Encounter*, edited by Martin Gren and Edward J. Huijbens, 34–52. London: Routledge, 2016.

Hurley, Frank. *Australia: A Camera Study*. Sydney, Australia: Angus and Robertson, 1955.

Hurley, Frank. *Pearls and Savages: Adventures in the Air, on Land and Sea—in New Guinea*. 1924; repr., New York: Putnam, 1925.

Hurley, Frank. *Queensland: A Camera Study*. Sydney: Angus and Robertson, 1950.

Jay, Martin. *Downcast Eyes: The Denigration of Vision in Twentieth-Century French Thought*. Berkeley: University of California Press, 1993.

Jeudy-Ballini, Monique. "Bringing Aesthetic Emotion into the Museum: Reflections on an Example from Oceania." THEMA: *La revue des musées de la civilisation* 2 (2015): 55–64.

Jolly, Martyn. "Australian First World War Photography: Frank Hurley and Charles Bean." *History of Photography* 23, no. 2 (1999): 141–48.

Kachur, Lewis. *Displaying the Marvelous: Marcel Duchamp, Salvador Dalí, and Surrealist Exhibition Installations*. Cambridge, MA: MIT Press, 2001.

Klein, Bernhard. Review of *The Ocean at Home: An Illustrated History of the Aquarium, and: Fathoming the Ocean: The Discovery and Exploration of the Deep Sea*, by Bernd Brunner. *Victorian Studies* 48, no. 4 (2006): 709–11.

Kort, Pamela. "Arnold Böcklin, Max Ernst, and the Debate around Origins and Survivals in Germany and France." In *Darwin: Art and the Search for Origins*, edited by Pamela Kort and Max Hollein, 24–92. Cologne, Germany: Wienand, 2009.

Krauss, Rosalind. "The Photographic Conditions of Surrealism." In *The Original-ity of the Avant-Garde and Other Modernist Myths*, 87–119. Cambridge, MA: MIT Press, 1986.

Larsen, Egon. *Men under the Sea*. London: Phoenix House, 1955.

Latour, Bruno. *We Have Never Been Modern*. Translated by Catherine Porter. Cambridge, MA: Harvard University Press, 1993.

Le Dû-Blayo, Laurence. "Underwater Landscapes in Comic Books." In *Underwater Seascapes: From Geographical to Ecological Perspectives*, edited by Olivier Musard, Laurence Le Dû-Blayo, Patrice Francour, Jean-Pierre Beurier, Eric Feunteun, and Luc Talassinos, 119–35. Cham, Switzerland: Springer, 2014.

Leslie, Esther. *Walter Benjamin: Overpowering Conformism*. London: Pluto Press, 2000.

Lingis, Alphonso. "Animal Body, Inhuman Face." In Wolfe, *Zoontologies*, 165–83.

Lingis, Alphonso. *Excesses: Eros and Culture*. Albany: State University of New York Press, 1983.

Lingis, Alphonso. "The Rapture of the Deep." In Alphonso, *Excesses*, 1–17.

Lingis, Alphonso. "Translator's Preface." In Merleau-Ponty, *Visible and the Invisible*, xl–lvi.

Lord, Walter. *Peary to the Pole*. New York: Harper and Row, 1963.

Lovejoy, Arthur O. "'Nature' as Aesthetic Norm." *Modern Language Notes* 42, no. 7 (November 1927): 444–50.

Loxley, Diana. *Problematic Shores: The Literature of Islands*. London: Macmillan, 1990.

Marshall, Justin, and Sönke Johnsen. "Camouflage in Marine Fish." In *Animal Camouflage: Mechanisms and Function*, edited by Martin Stevens and Sami Merilaita, 186–282. Cambridge: Cambridge University Press, 2011.

Martínez, Alejandro. "A Souvenir of Undersea Landscapes: Underwater Photography and the Limits of Photographic Visibility 1890–1910." Translated by Catherine Jagoe. *História, Ciências, Saúde—Manguinhos, Rio de Janeiro* 21, no. 3 (2014), 1029–47. Accessed December 17, 2017. http://dx.doi.org/10.1590/S0104 -59702014000300013.

Marx, Robert F. *Into the Deep: The History of Man's Underwater Exploration*. New York: Van Nostrand Reinhold, 1978.

Matsuda, Matt K. *The Memory of the Modern*. New York: Oxford University Press, 1996.

McCalman, Iain. *The Reef: A Passionate History*. Melbourne, Australia: Viking, 2013.

McCalman, Iain. "Turtle War: Captain Cook's Environmental Crisis on the Great Barrier Reef." Vaughan Evans Memorial Lecture 2011. *Great Circle* 34, no. 2 (2012): 7–18.

McCulloch, Allan R. "Fishes and the Movies." *Australian Museum Magazine* 2, no. 3 (July–September 1924): 103–8.

McGranahan, Gordon, Deborah Balk, and Bridget Anderson. "The Rising Tide: Assessing the Risks of Climate Change and Human Settlements in Low Elevation Coastal Zones." *Environment and Urbanization* 19, no. 1 (2007): 17–37.

McGregor, Alasdair. *Frank Hurley: A Photographer's Life.* Camberwell, Australia: Viking, 2004.

McMahon, Thomas J. "The Pearl Fishers of Torres Strait Islands." *Geographical Review* 10, no. 3 (September 1920): 182–84.

McNab, Robert. *Ghost Ships: A Surrealist Love Triangle.* New Haven, CT: Yale University Press, 2004.

McNeill, J. R., and Peter Engelke. *The Great Acceleration: An Environmental History of the Anthropocene since 1945.* Cambridge, MA: Belknap Press of Harvard University Press, 2014.

McNiven, Liz. "Curator's notes." *National Film and Sound Archives.* Accessed 14 June, 2018. https://aso.gov.au/titles/documentaries/pearls–and–savages/notes/.

Mentz, Steve. *At the Bottom of Shakespeare's Ocean.* London: Continuum, 2009.

Mentz, Steve, and Martha Elena Rojas, eds. "Introduction: 'The Hungry Ocean.'" In *The Sea and Nineteenth-Century Anglophone Literary Culture,* 1–15. Abingdon, UK: Routledge, 2017.

Merleau-Ponty, Maurice. *The Visible and the Invisible.* Translated by Alphonso Lingis. Evanston, IL: Northwestern University Press, 1968.

Merleau-Ponty, Maurice. "Working Notes." In Merleau-Ponty, *Visible and the Invisible,* 165–277.

Michelet, Jules. *The Sea.* New York: Rudd and Carleton, 1861.

Millar, David P. *From Snowdrift to Shellfire: Capt. James Francis (Frank) Hurley (1885–1962).* Sydney, Australia: David Ell Press, 1984.

Miller, John. *Empire and the Animal Body: Violence, Identity, and Ecology in Victorian Adventure Fiction.* London: Anthem Press, 2014.

Miner, Roy Waldo. "Diving in Coral Gardens: A Scientist Works beneath the Clear Waters of the Coral Reefs of the Bahamas." *Scientific American* 151, no. 3 (1934): 122–24.

Mitchell, W. J. T. *What Do Pictures Want? The Lives and Loves of Images.* Chicago: University of Chicago Press, 2005.

Mitman, Gregg. "Cinematic Nature: Hollywood Technology, Popular Culture, and the American Museum of Natural History." *History of Science Society* 84, no. 4 (December 1993): 637–61.

Mitman, Gregg. *Reel Nature: America's Romance with Wildlife on Film.* Cambridge, MA: Harvard University Press, 1999.

Moore, Clive. "Empires of the Coral Sea." In *The Routledge History of Western Empires,* edited by Robert Aldrich and Dirsten McKenzie, 151–65. London: Routledge, 2014.

Morrison, Toni. *The Origin of Others.* Cambridge, MA: Harvard University Press, 2017.

Morton, Timothy. *The Ecological Thought*. Cambridge, MA: Harvard University Press, 2010.

Morton, Timothy. *Ecology without Nature: Rethinking Environmental Aesthetics*. Cambridge, MA: Harvard University Press, 2007.

Morton, Timothy. "X-Ray." In Cohen, *Prismatic Ecology*, 311–27.

Napier, Sydney Elliott. *On the Barrier Reef: Notes from a No-ologist's Pocket-Book*. Sydney: Angus and Robertson, 1939. Originally published 1928.

Nemerov, Alexander. "Vanishing Americans: Abbott Thayer, Theodore Roosevelt, and the Attraction of Camouflage." *American Art* 11, no. 2 (1997): 50–81.

Norton, Trevor. *Stars beneath the Sea: The Extraordinary Lives of the Pioneers of Diving*. London: Century, 1999.

Olcott, Frances Jenkins. *Our Wonderful World*. Boston, MA: Little, Brown, 1936.

Orams, Mark. *Marine Tourism: Development, Impacts, and Management*. London: Routledge, 1999.

Osgerby, Bill. "Rapture of the Deep: Leisure, Lifestyle and the Lure of Sixties Scuba." In *Historicizing Lifestyle: Mediating Taste, Consumption, and Identity from the 1900s to 1970s*, edited by David Bell and Joanne Hollows, 88–108. London: Routledge, 2006.

Papers of Frank Hurley. MS883. National Library of Australia, Canberra, Australia.

Parkinson, Gavin. "Emotional Fusion with the Animal Kingdom." In *The Art of Evolution: Darwin, Darwinisms, and Visual Culture*, edited by Barbara Larson and Fae Brauer, 262–88. Hanover, NH: Dartmouth College Press, 2009.

Peck, Robert McCracken. "The Explorers Club." *Magazine Antiques* 166, no. 6 (December 2004): 96–101.

Petterson, Palle B. *Cameras into the Wild: A History of Early Wildlife and Expedition Filmmaking, 1895–1928*. Jefferson, NC: McFarland, 2011.

Picard, David. "Island Magic: Tourism and the Dialectics of Self-imaging in La Reunion." In *Experiencing Diversity and Mutuality, 10th Biennial Conference of EASA (European Association of Social Anthropologists)*, Ljubljana, Slovenia, August 26–29, 2008, 26–30.

Pike, A. F. "Hurley, James Francis (Frank) (1885–1962)." *Australian Dictionary of Biography*. National Centre of Biography, Australian National University. Published first in hardcopy in 1983. Accessed online February 22, 2018. http://adb.anu.edu.au/biography/hurley-james-francis-frank-6774/text11715.

Plimpton, George, ed. *As Told at The Explorers Club: More than Fifty Gripping Tales of Adventure*. Guilford, CT: Lyons Press, 2003.

Pocock, Celmara. "Aborigines, Islanders and Hula Girls in Great Barrier Reef Tourism." *Journal of Pacific History* 49, no. 2 (2014): 170–92.

Pocock, Celmara. "Entwined Histories: Photography and Tourism at the Great Barrier Reef." In *The Framed World: Tourism, Tourists and Photography*, edited by Mike Robinson and David Picard, 185–97. Farnham, UK: Ashgate Publishing, 2009.

Polizzotti, Mark. *Revolution of the Mind: The Life of André Breton*. London: Bloomsbury Publishing, 1995.

Porter, Roy. "Baudrillard: History, Hysteria and Consumption." In *Forget Baudrillard?*, edited by Chris Rojek and Bryan S. Turner, 1–22. London: Routledge, 1993.

Putnam, G. P. Foreword to *Pearls and Savages: Adventures in the Air, on Land and Sea—in New Guinea*, by Frank Hurley. 1924, repr., New York: Putnam, 1925.

Quinn, Stephen Christopher. *Windows on Nature: The Great Habitat Dioramas of the American Museum of Natural History*. New York: Abrams, in association with the American Museum of Natural History, 2006.

Rancière, Jacques. *The Politics of Aesthetics*. Translated and introduction by Gabriel Rockhill. London: Continuum, 2007.

Rediker, Marcus. "History from below the Water Line: Sharks and the Atlantic Slave Trade." *Atlantic Studies* 5, no. 3 (2008): 285–97.

Reidy, Michael S., Gary Kroll, and Erik M. Conway. *Exploration and Science: Social Impact and Interaction*. Santa Barbara, CA: AB-CLIO, 2007.

Rice, Alan. "Food for the Sharks." In *Radical Narratives of the Black Atlantic*, 48–82. London: Continuum, 2003.

Robins, Kevin. *Into the Image: Culture and Politics in the Field of Vision*. London: Routledge, 1996.

Robins, William D., Mizue Hisano, Sean R. Connolly, and J. Howard Choat. "Ongoing Collapse of Coral-Reef Shark Populations." *Current Biology* 16 (2006): 2314–19.

Rohman, Carrie. *Stalking the Subject: Modernism and the Animal*. New York: Columbia University Press, 2009.

Roosevelt, Theodore. "Nature Fakers." (1907) In *Roosevelt's Writing: Selections from the Writings of Theodore Roosevelt*, edited by Maurice G. Fulton. New York: Macmillan, 1920.

Rozwadowski, Helen M. *Fathoming the Ocean: The Discovery and Exploration of the Deep Sea*. Foreword by Sylvia A. Earle. Cambridge, MA: Belknap Press of Harvard University Press, 2005.

Rozwadowski, Helen M. "Oceans: Fusing the History of Science and Technology with Environmental History." In *A Companion to American Environmental History*, edited by Douglas Cazaux Sackman, 442–61. Chichester, UK: Wiley-Blackwell, 2010.

Rozwadowski, Helen M. "Playing by—and on and under the Sea: The Importance of Play for Knowing the Ocean." In *Knowing Global Environments: New Historical Perspectives on the Field Sciences*, edited by Jeremy Vetter, 162–90. New Brunswick, NJ: Rutgers University Press, 2011.

Rozwadowski, Helen M., and David K. van Keuren, eds. Introduction to *The Machine in Neptune's Garden: Historical Perspectives on Technology and the Marine Environment*, xiii–1. Sagamore Beach, MA: Science History Publications, 2004.

Ryan, James R. *Photography and Exploration*. London: Reaktion Books Exposures Series, 2013.

Ryan, James R. *Picturing Empire: Photography and the Visualization of the British Empire*. London: Reaktion Books, 1997.

Ryan, Simon. *The Cartographic Eye: How Explorers Saw Australia*. Cambridge: Cambridge University Press, 1996.

Said, Edward. *Culture and Imperialism*. London: Chatto and Windus, 1993.

Saler, Michael T. "Modernity, Disenchantment, and the Ironic Imagination." *Philosophy and Literature* 28, no. 1 (2004): 137–49.

Sanitarium Health Food Company. *Marvels of the Great Barrier Reef: A Pictorial Record of Australia's Marine Wonderland*. Sydney, Australia: Sanitarium Health Food Company, 1948.

Saunders, Brian. *Discovery of Australia's Fishes: A History of Australian Ichthyology to 1930*. Collingwood, Australia: CSIRO Publishing, 2012.

Saville-Kent, William. *The Great Barrier Reef of Australia, Its Products and Potentialities*. London: W. H. Allen, 1893.

Schwartz, Dona. "Visual Ethnography: Using Photography in Qualitative Research." *Qualitative Sociology* 12, no. 2 (1989): 119–54.

Schwartz, Herbert F. "In Brightest Africa." *Natural History: The Journal of the American Museum of Natural History* 24 (1924): 109–11.

Schwartz, Hillel. *The Culture of the Copy: Striking Likenesses, Unreasonable Facsimiles*. New York: Zone Books, 1996.

Seaman, William, Jr., and Anthony C. Jensen. "Purposes and Practices of Artificial Reef Evaluation." In *Artificial Reef Evaluation: With Application to Natural Marine Habitats*, edited by William Seaman, Jr., 1–21. London: Taylor and Francis, 2000.

Seelye, John. "Oceans of Emotion: The Narcissus Syndrome." *Virginia Quarterly Review* 58, no. 2 (spring 1982): 189–207.

Sekula, Allan. *Fish Story: Witte de With Center for Contemporary Art, Rotterdam, 21.01.1995–12.03.1995*. Düsseldorf, Germany: Richter Verlag, 1995.

Sharp, Nonie. *No Ordinary Judgment: Mabo, the Murray Islanders' Land Case*. Canberra, Australia: Aboriginal Studies Press, 1996.

Sheller, Mimi. "Demobilizing and Remobilizing Caribbean Paradise." In Sheller and Urry, *Tourism Mobilities*, 14–20.

Sheller, Mimi, and John Urry, eds. *Tourism Mobilities: Places to Play, Places in Play*. London: Routledge, 2004.

Sherwood, George F. "All in the Day's Work." In Blossom, *Told at the Explorer's Club*, 277–95.

Shick, J. Malcolm. "Toward an Aesthetic Marine Biology." *Art Journal* 67, no. 4 (2008): 62–86.

Soister, John T., Henry Nicolella, Steve Joyce, and Harry N. Long. "20,000 Leagues Under the Sea." In *American Silent Horror, Science Fiction, and Fantasy Films, 1913–1929*, 587–92. Jefferson, NC: McFarland, 2012.

Sontag, Susan. *On Photography.* New York: Farrar, Straus and Giroux, 1977; repr., Auckland: Penguin, 1982.

Soper, Kate. *What Is Nature?: Culture, Politics, and the Non-human.* Oxford: Blackwell, 1995.

Specht, Jim, and John Fields. *Frank Hurley in Papua: Photographs of the 1920–1923 Expeditions.* Bathurst, Australia: Robert Brown in association with the Australian Museum Trust, 1984.

Sponsel, Alistair. "Darwin's Theory of Coral Reef Formation." In *Coral: Something Rich and Strange,* edited by Marion Endt-Jones, 70–75. Liverpool, UK: Liverpool University Press, 2013.

Starosielski, Nicole. "Beyond Fluidity: A Cultural History of Cinema under Water." In *Ecocinema Theory and Practice,* edited by Stephen Rust, Salma Monani, and Sean Cubitt, 149–68. New York: Routledge, 2013.

Stephen, Ann, Andrew McNamara, and Philip Goad, eds. *Modernism and Australia: Documents on Art, Design, and Architecture 1917–1967.* Carlton, Australia: Miegunyah Press, 2006.

Stephenson, Ellen Mary, and Charles Samuel Stewart. *Animal Camouflage.* Harmondsworth, UK: Penguin Books, 1946.

Sykes, A. Krista. "Underwater Photography." In *Encyclopedia of Twentieth Century Photography,* edited by Lynne Warren, 1581–84. New York: Routledge, 2006.

Taves, Brian. *Hollywood Presents Jules Verne: The Father of Science Fiction on Screen.* Lexington: University Press of Kentucky, 2015.

Taves, Brian. "A Pioneer under the Sea: Library Restores Rare Film Footage." *Information Bulletin* (Library of Congress) 55, no. 15 (September 16, 1996). Accessed December 1, 2017. http://www.loc.gov/loc/lcib/9615/sea.html.

Taves, Brian. "With Williamson beneath the Sea." *Journal of Film Preservation* 25, no. 52 (1996): 54–61. Accessed December 21, 2017. http://jv.gilead.org.il/taves/withwill.html.

Thanhouser, Ned. "In the Tropical Seas (1914)." Vimeo video, 13:14. Released September 1914. Posted March 16, 2011. https://vimeo.com/21137464.

Thayer, Abbott H. "The Law Which Underlies Protective Coloration." *Auk* 13, no. 2 (April 1896): 124–29.

Thompson, Krista A. *An Eye for the Tropics: Tourism, Photography, and Framing the Caribbean Picturesque.* Durham, NC: Duke University Press, 2006.

Todd, David. *Astronomy: The Science of the Heavenly Bodies.* New York: Harper and Brothers, 1922.

Torma, Franziska. "Frontiers of Visibility: On Diving Mobility in Underwater Films (1920s to 1970s)." *Transfers* 3, no. 2 (2013): 24–46.

Torre, Stephen. "Tropical Island Imaginary." *Etropic: Tropics of the Imagination 2013 Proceedings* 12, no. 2 (2013): 246–73.

Toscano, Alberto, and Jeff Kinkle. *Cartographies of the Absolute.* Winchester, UK: Zero Books, 2015.

Tucker, Jennifer. "The Historian, the Picture, and the Archive." *Isis* 97, no. 1 (2006): 111–20.

Urry, John. *The Tourist Gaze; Leisure and Travel in Contemporary Societies.* London: Sage, 1990.

Urry, John. "The Tourist Gaze 'Revisited.'" *American Behavioral Scientist* 36, no. 2 (1992): 172–86.

Van Liere, Eldon N. "On the Brink: The Artist and the Sea." In *Poetics of the Elements in the Human Condition: The Sea,* edited by Anna-Teresa Tymieniecka, 269–86. Dordrecht: Springer Netherlands, 1985.

Verne, Jules. *Twenty Thousand Leagues Under the Sea: An Underwater Tour of the World.* Translated by F. P. Walter. Reprint of the 1992 translation by F. P. Walter of the original 1869–71 edition published by Paris by J. Hetzel et Cie. Project Gutenberg, 2002. http://www.gutenberg.org/cache/epub/2488/pg2488.txt.

Veron, J. E. N. *A Reef in Time: The Great Barrier Reef from Beginning to End.* Cambridge, MA: Belknap Press of Harvard University Press, 2009.

Vesely, Dalibor. "Surrealism, Myth and Modernity." *Architectural Design* 48, no. 2–3 (1978): 86–95.

Vidler, Anthony. *The Architectural Uncanny: Essays in the Modern Unhomely.* Cambridge, MA: MIT Press, 1992.

Von Moltke, Johannes. "Ruin Cinema." In *Ruins of Modernity,* edited by Julia Hell and Andreas Schönle, 395–417. Durham, NC: Duke University Press, 2010.

Warlick, M. E. "Magic, Alchemy and Surrealist Objects." In *Magical Objects: Things and Beyond (Leipzig Explorations in Literature and Culture 12),* edited by Elmar Schenkel and Stefan Welz, 1–33. Glienicke, Germany: Galda + Wilch Verlag, 2007.

Warner, Marina. "Here Be Monsters." *New York Review of Books,* December 19, 2013–January 8, 2014, 58–62.

Webb, Virginia-Lee. "Official/Unofficial Images: Photographs from the Crane Pacific Expedition, 1928–1929." *Pacific Studies* 20, no. 4 (1997): 103–24.

Weber, Samuel M. "The Madrepore." *Modern Languages Notes* 87, no. 7 (1972): 915–61.

Westell, William Percival. "Monsters of the Mussel Family." In Hammerton, *Our Wonderful World,* 2:882.

White, Lynn, Jr. "The Historical Roots of our Ecological Crisis." In *Ecology and Religion in History,* edited by David Spring and Eileen Spring, 15–32. New York: Harper and Row, 1974.

Whitley, David. "Animation, Realism, and the Genre of Nature." In *Moving Environments: Affect, Emotion, Ecology, and Film,* edited by Alexa Weik von Mossner, 143–59. Waterloo, ON: Wilfrid Laureier University Press, 2014.

Wijgerde, Tim. "Inside a Coral Lab." *Advanced Aquarist* 10 (September 2011). Accessed December 20, 2017. http://www.advancedaquarist.com/2011/9/aafeature.

Williamson, John Ernest. *Twenty Years under the Sea*. Boston, MA: Hale, Cushman and Flint, 1936.

Williamson, John Ernest. "Under the Sea, 1929 (Reel 1)." Internet Archive video, 15:22. Posted December 2, 2009 by Field Museum of Natural History. https://archive.org/details/UnderTheSeareel1.

Williamson, John Ernest, and Frances Jenkins Olcott. *Child of the Deep*. Boston, MA: Houghton Mifflin Company, 1938.

Winton, Tim. *The Boy behind the Curtain*. Melbourne, Australia: Hamish Hamilton (Penguin Books and Random House), 2016.

Withers, Karen J. T. "Empire Building Colonials: The Implications of Size in the Hard Corals." PhD diss., University of Sydney, 2000. https://ses.library.usyd.edu.au/handle/2123/9033.

Wolfe, Cary. *Zoontologies: The Question of the Animal*. Minneapolis: University of Minnesota Press, 2003.

Woods, Peter H. *Weathering the Storm: Inside Winslow Homer's Gulf Stream*. Athens: University of Georgia Press, 2004.

Wulf, Andrea. *The Invention of Nature: The Adventures of Alexander von Humboldt, the Lost Hero of Science*. London: John Murray, 2015.

INDEX

Italic page numbers indicate images.

Adventures on the Ocean Floor, 107
l'Aquarium Humain, 33
affect, 218: coral reef animals and, 23, 217–18, 222; nature and, 224
African-Bahamian, 66, 84: Williamson and, 42, 56, 80–81, 86–88, 44, 57. See also Bahamas; divers; labor; spectacle; racism
Akeley, Carl, 11, 75, 185–86, 191–96
animals, 10, 49: animal behavior, 161; camouflage and, 22-23, 64-65, 161–62, 212, 217–218; clams, giant, 109, 169; crittercam, 227–28; cinema, 88, 92–93, 102; mythical. See monsters; octopuses, 43, 63, 72, 101: mechanical, 193–94; objectification of, 1, 52, 65, 109, 219–20, 234; photography of 153–54, 159, 160; human-animal relationship ("political ecology," cohabitation), 219–20, 227–28; as victims of physical violence, 9, 73, 74, 75, 79–80, 83, 89–93, 140. See also camouflage; fish; sharks
Anthropocene, the, 218, 222
aqualung, 198, 206, 211–12
aquariums: aquarium thinking, 6, 33, 126–27, 206, 232; Australia, as symbols of modernity, 126; contemporary art and, 225–26; as a diorama, 129, 156; British Empire and, 52, 144; cam-

ouflage and, 142–43, 161–62, 285; colonialism, as symbol of, 135; framing effect of, 36, 52, 151, 154–55, 156. See also constructed reality; photosphere; glass windows, paradoxical effect of and, 6, 52, 146, 152–53, 232. See also glass windows, paradoxical effect of; photography, 33, 128, 151–52, 159, 161. See also constructed reality; photosphere; spectatorship
Australia: André Breton and, 41; aquariums, as symbols of modernity, 126; colonization, 8; coral reefs, traditional ownership of, 122; popular press and, 97–101; the photosphere and, 100–101, 110–11; sharks and monsters and, 101–2, 104, 108–9; "tropical possessions" (Australian Commonwealth administered regions), 119; Williamson and, 96, 99; Williamson, as a hero, 95–100
Australia: A Camera Study (1955), 202–9
avant-garde. See surrealism

Bahamas, 65: black labor and pearl diving 171–72; colonization, 8, 19, 84–85; pearl diving, 172; quality of light, 65–66; racism and, 19, 56, 84–86; sharks in cultural memory, 85; tourism and, 67, 204

www.ingramcontent.com/pod-product-compliance
Lightning Source LLC
Chambersburg PA
CBHW051210170526
45166CB00005B/1826